AQA
A-level

Further
Mathematics
Discrete

Author
Nick Geere

Series editor
Heather Davis

HODDER
EDUCATION
AN HACHETTE UK COMPANY

Orders: please contact Hachette UK Distribution, Hely Hutchinson Centre, Milton Road, Didcot, Oxfordshire, OX11 7HH. Telephone: +44 (0)1235 827827. Email education@hachette.co.uk Lines are open from 9 a.m. to 5 p.m., Monday to Friday. You can also order through our website: www.hoddereducation.co.uk

© Nick Geere 2018

Published by Hodder Education

An Hachette UK Company

Carmelite House, 50 Victoria Embankment, London EC4Y 0DZ

Impression number   5

Year   2024

Cover photo © Rachael Arnott/stock.adobe.com

Illustrations by Aptara Inc.

Typeset in bemboStd 11/13 pts. by Aptara Inc.

Printed and bound by CPI Group (UK) Ltd, Croydon, CR0 4YY

A catalogue record for this title is available from the British Library

ISBN 9781510433342

Contents

Getting the most from this book

Mathematics is not only a beautiful and exciting subject in its own right but also one that underpins many other branches of learning. It is consequently fundamental to our national wellbeing.

This book covers the Discrete Mathematics elements in the AQA AS and A Level Further Mathematics specifications. Students start these courses at a variety of stages. Some embark on AS Further Mathematics in Year 12, straight after GCSE, taking it alongside AS Mathematics, and so have no prior experience of A Level Mathematics. In contrast, others only begin Further Mathematics when they have completed the full A Level Mathematics course. This book requires no prior knowledge of A level Mathematics and so can be started at any time. The last chapter requires the manipulation of matrices, covered in *AQA A Level Further Mathematics Core Year 1 (AS)* but earlier work using matrices is accessible from GCSE. The last chapter also refers briefly to the multiplication of imaginary numbers, but again, this can be accessed before covering them in depth.

Between 2014 and 2016 A Level Mathematics and Further Mathematics were very substantially revised, for first teaching in 2017. Changes that particularly affect Discrete Mathematics include increased emphasis on

- Problem solving
- Mathematical rigour
- Use of ICT
- Modelling.

This book embraces these ideas. A large number of exercise questions involve elements of problem solving and require the application of the ideas and techniques in a wide variety of real world contexts. This develops independent thinking and builds on thorough understanding. Discrete Mathematics often provides descriptions of real world situations that make them tractable to calculations, and so modelling is key to this branch of mathematics. It pervades much of the book, particularly the chapters on the use of graphs to solve real world problems.

Throughout the book the emphasis is on understanding and interpretation rather than mere routine calculations, but the various exercises do nonetheless provide plenty of scope for practising basic techniques. The exercise questions are split into three bands. Band 1 questions (indicated by a light grey) are designed to reinforce basic understanding; Band 2 questions (a darker bar) extend the difficulty level for more in-depth understanding; Band 3 questions (a darker bar again) explore around the topic and some of them are rather more demanding. Where appropriate, there is a fuller exercise included at the end of the Year 1 material or at the end of the chapter after the Year 2 material. The final exercise in each chapter starts with questions on the content of the last section and then may include some more demanding synoptic questions. In addition, extensive online support, including further questions, is available by subscription to MEI's Integral website, http://integralmaths.org.

At the end of each chapter there is a list of key points covered as well as a summary of the new knowledge (learning outcomes) that readers should have gained.

Two common features of the book are Activities and Discussion points. These serve rather different purposes. The Activities are designed to help readers get into the thought processes of the new work that they are about to meet; having done an Activity, what follows will seem much easier. The Discussion points invite readers to talk about particular points with their fellow students and their teacher and so enhance their understanding.

Answers to all exercise questions are provided at the back of the book, and also online at www.hoddereducation.co.uk/AQAFurtherMathsDiscrete.

*Please note that mark allocations given in assessment questions are to be used as guidelines only: AQA have not reviewed or approved these marks.

Prior knowledge

No prior knowledge of discrete mathematics is needed for this book. It does, however, assume that the reader is reasonably fluent in basic algebra and graphs: working with formulae and expressions; solving linear simultaneous equations; graphing inequalities.

Matrices are used to store information and later, 2 by 2 matrices are multiplied as members of a group. Familiarity with the scalar product of vectors and working with imaginary numbers are also helpful for the work on group theory.

Chapter 1 Graphs

This chapter introduces the basic ideas and vocabulary associated with graphs in discrete mathematics. Matrices are mentioned but no prior knowledge is needed to access the work involving them. The Year 2 work in Sections 1.4 and 1.5 requires knowledge of the earlier sections.

Chapter 2 Networks

Chapter 2 builds on the work in Chapter 1, adding weights to the edges, now referred to as arcs. More vocabulary is introduced.

Chapter 3 Linear programming

This builds on the work at GCSE on linear graphs and graphing linear inequalities in two variables. Solving linear simultaneous equations is also required from GCSE. Work on the Simplex algorithm requires knowledge of the Year 1 work in the first two sections.

Chapter 4 Critical path analysis

This builds on the work in Chapters 1 and 2 but does not depend on it. The Year 2 work in Sections 4.4 and 4.5 depends on the work in the earlier sections.

Chapter 5 Network flows

This builds on the work in Chapters 1 and 2. The Year 2 work in Sections 5.4 and 5.5 depends on the work in the earlier sections.

Chapter 6 Game theory

This uses work on linear equations from GCSE and uses work on the Simplex algorithm from Chapter 3. The Year 2 work on converting to linear programming problems also requires knowledge of the earlier sections in this chapter.

Chapter 7 Binary operations and group theory

This uses work from Further Pure Mathematics on the scalar product of vectors and multiplication of 2 by 2 matrices and imaginary numbers. The Year 1 work on binary operations is required for the remainder of the chapter.

1

Graphs

→ Can you draw this diagram without lifting your pen from the paper or repeating any line?

1 The language of graphs

The types of graph that are considered here are different from the graphs of functions that you will be familiar with. For example, the graph in Figure 1.1 shows connections between pairs of members of a discrete set (of towns and cities in south-west England). The connections could, for example, represent the existence of a direct bus route.

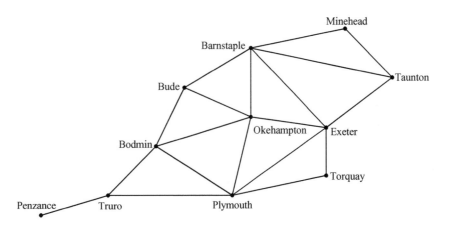

Figure 1.1

This chapter introduces the main terminology that is used in graph theory. It will be needed in subsequent chapters, where some of the applications of the theory are explored.

In Figure 1.1, the towns (and cities) are referred to as **vertices**, whilst the connections between them are described as **edges**. An edge must have a vertex at each end.

In Chapter 2 you will meet **networks**. These are graphs with numbers associated with the edges, called weights (e.g. distances, travel times, costs).

In Chapter 5 (Network flows), you will work with **directed graphs** (or digraphs), where an edge may have a direction associated with it.

Another example of a graph is shown in Figure 1.2. It represents the relationship 'share a common factor other than 1'.

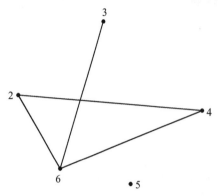

Figure 1.2

- Within a particular graph, a **walk** is a sequence of edges in which the end of one edge is the start of the next (except for the last edge).

- A **trail** is a walk in which no edge is repeated (but you are allowed to pass through a particular vertex more than once).

- A trail that starts and ends at the same vertex is a **closed trail**.

- A **path** is a trail with the further restriction that no vertex is repeated.

- A closed path is called a **cycle**.

Example 1.1

Look at Figure 1.1. What description would you give to the following routes?

(i) Okehampton–Exeter–Barnstaple–Okehampton–Plymouth–Bodmin–Okehampton

(ii) Okehampton–Bodmin–Truro–Plymouth–Torquay

Solution

(i) It is at least a walk, as each edge follows the previous one.
It is at least a trail, as no edge is repeated.
It is a closed trail, as it returns to its starting point at Okehampton.
It is not a path, as the vertex Okehampton is repeated.
So it is a closed trail.

(ii) It is at least a walk, as each edge follows the previous one.
It is at least a trail, as no edge is repeated.
It is not closed, as it does not return to its starting point at Okehampton.
It is at least a path, as no vertex is repeated.
So it is a path.

- A graph is said to be **connected** if there exists a path between every pair of vertices; i.e. if no vertices are isolated. This means that the graph in Figure 1.2 is not connected.

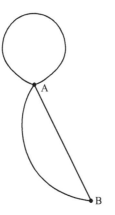

- It is possible for two vertices to be connected by **multiple edges**, or for a vertex to be connected to itself (forming a **loop**). These situations are shown in Figure 1.3.

Figure 1.3

- A graph that has no multiple edges or loops is referred to as a **simple graph**.

- A **tree** is a simple connected graph with no cycles. Figure 1.4 shows an example.

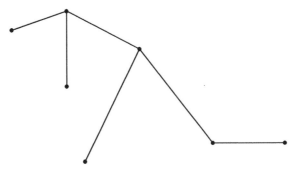

Discussion point

➜ What is the smallest number of edges that a simple connected graph with *n* vertices can have?

Figure 1.4

- A **subgraph** is a graph that is formed from some of the vertices and edges of another graph. Note that whilst this may result in an isolated vertex, any edge has to have a vertex at each end.

- A subgraph *H*, of a connected graph *G*, is said to be a **spanning tree** of *G* if *H* is a tree and it contains all the vertices of *G*. One particular spanning tree for the graph in Figure 1.1 is shown in Figure 1.5. In general, there may be many possible spanning trees.

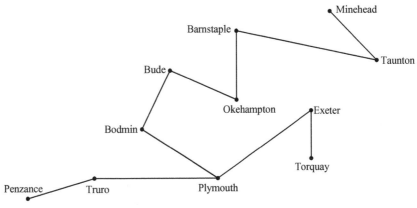

Figure 1.5

This concept will be of use in Chapter 2, where you will consider the minimum connector problem.

■ The **degree** (or sometimes **order**) of a vertex is the number of edges that join it. (A loop contributes two to the degree of its vertex.) A vertex that has an odd degree, for example, can be referred to as an odd vertex.

ACTIVITY 1.1

Prove that, in a graph, the number of odd vertices is always even.

Exercise 1.1

① (i) List all the cycles in the graph below that can start and finish at A. Note that, for example, ABCA and ACBA represent the same cycle.

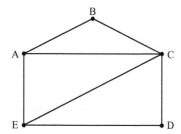

Figure 1.6

(ii) Why is ABCEDCA not a cycle?

(iii) What could ABCEDCA best be described as?

② Draw three different trees, each containing five vertices and four edges. [MEI]

③ Vertices of the graph shown in Figure 1.7 represent objects. Some edges have been drawn to connect vertices representing objects which are the same colour.

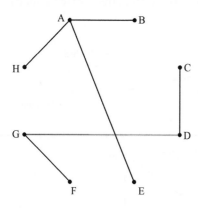

Figure 1.7

(i) Copy the diagram and draw in whichever edges you can be sure should be added.

(ii) How many edges would be needed in total if you were also told that the objects represented by B and F were the same colour? [MEI]

④ Draw a graph that is not connected and not simple, and has six vertices, all of degree 3.

⑤ Table 1.1 shows the numbers of vertices of degrees 1, 2, 3 and 4 in four different graphs. Draw an example of each of these graphs.

Degree of vertex	1	2	3	4
Graph 1	4	0	0	1
Graph 2	0	0	4	1
Graph 3	0	1	0	1
Graph 4	2	0	0	1

Table 1.1 [MEI]

⑥ A particular simple connected graph has five vertices and seven edges, and the degree of each vertex is either 2, 3 or 4.

(i) Explain why the sum of the degrees of the vertices is 14.

(ii) Copy and complete Table 1.2 to show two of the possibilities for the numbers of vertices of each degree.

Number of vertices	Number of degree 2	Number of degree 3	Number of degree 4	Sum of degrees
5				14
5				14

Table 1.2

(iii) Draw a diagram for each of your possibilities from part (ii). [MEI]

⑦ (i) A simple connected graph has seven vertices, all having the same degree d. Give the possible values of d, and for each value of d give the number of edges of the graph.

(ii) Another simple connected graph has eight vertices, all having the same degree d.

Draw such a graph with $d = 3$, and give the other possible values of d.

(iii) Explain why there are no odd values for d in (i) and why it is possible for d to be odd in (ii).

2 Eulerian, Hamiltonian and planar graphs

Eulerian graphs

An interesting problem is whether it is possible to travel round a graph without repeating any edges (in other words, along a trail), so that all the edges in the graph are covered.

> If such a trail ends at its starting point, it is called **Eulerian**. If it ends somewhere else, it is called **semi-Eulerian**.

A graph that possesses an Eulerian or semi-Eulerian trail is called an Eulerian or semi-Eulerian graph, as appropriate.

A practical example of this would be a gritting lorry that needs to travel down all the roads in a particular area, without repeating any roads if it can. (It is assumed that the roads are narrow, so the lorry need travel in one direction only.) In the case of an Eulerian graph, the lorry would be able to return to its depot, whereas in the case of a semi-Eulerian graph it would not. This is an example of the 'route inspection problem' that you will meet in Chapter 2.

> **Historical note**
>
> Leonhard Euler (1707–1783) was a very distinguished and versatile Swiss mathematician. His name is pronounced 'oiler'.

> If a graph has no odd vertices, then it can be shown to be Eulerian.

> You saw earlier in Activity 1.1 that the number of odd vertices of a graph is always even.

For every edge leading into a vertex, there will be another edge leading out, and it will be possible to move round the graph, covering each edge exactly once, without getting stranded at any vertex.

> If just two of the vertices are odd, then the graph can be shown to be semi-Eulerian.

You can start at one of these odd vertices, cover all of the edges exactly once, and end up at the other odd vertex.

> **Note**
>
> For the start vertex, the number of outgoing edges is one greater than the number of incoming edges (and the other way round for the end vertex).

Hamiltonian graphs

Another interesting problem is finding a route around a graph that visits all of the vertices exactly once. Note that edges cannot be repeated, as this would mean the repetition of a vertex (however, not all edges need to be traversed). You also need to be able to return to the starting point. If such a route exists, then it is called a **Hamiltonian cycle** (also known as a tour), and a graph that possesses a Hamiltonian cycle is called a Hamiltonian graph. This idea is employed in the 'Travelling salesperson' problem that you will meet in Chapter 2.

ACTIVITY 1.2
Look back at Figure 1.1 again. Which town or city should be removed, in order to make the graph Hamiltonian?

Planar graphs

A graph is said to be **planar** if it can be distorted in such a way that its edges do not cross. Figure 1.8 is planar, because it can be redrawn as Figure 1.9.

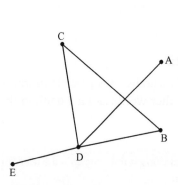

Figure 1.8

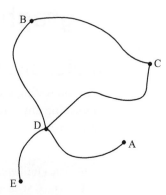

Figure 1.9

If a graph is both connected and planar, and if you consider a representation of the graph where the edges don't cross, then the plane containing the graph can be divided up into faces (or regions), which are bounded by the edges. You also include the infinite face with no boundary.

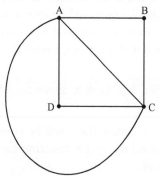

Figure 1.10

So the graph in Figure 1.10 has 4 faces, while Figure 1.9 has 2 faces.

Euler found that the following result holds for all connected planar graphs.

$$F + V = E + 2$$

ACTIVITY 1.3
Verify Euler's formula for the graph in Figure 1.1.

where F, V and E are the numbers of faces, vertices and edges, respectively. This result is known as **Euler's formula**.

Example 1.2

Use Euler's formula to show that if the number of vertices in a connected planar graph is one greater than the number of edges, then the graph is a tree.

Solution

From Euler's formula, $F + V = E + 2$.

Then, if $V = E + 1$, $F = 1$.

If there is only one face, then there are no cycles, multiple edges or loops, and so the graph is a simple connected graph with no cycles; i.e. a tree.

Example 1.3

(i) Find a Hamiltonian cycle for the graph below.

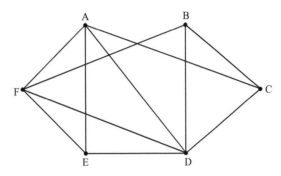

Figure 1.11

(ii) Starting with your cycle, add in edges to show that the original graph is planar.

Solution

(i) A Hamiltonian cycle is shown in Figure 1.12.

(ii) For example, you can start with the cycle shown in Figure 1.12, and manoeuvre vertices B and C, in order to remove crossing edges (giving Figure 1.13). You then add in the necessary (non-crossing) edges to give Figure 1.14, which is a distorted version of the original graph, showing that the original graph is planar.

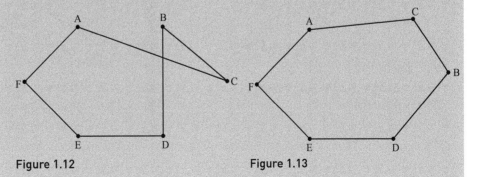

Figure 1.12

Figure 1.13

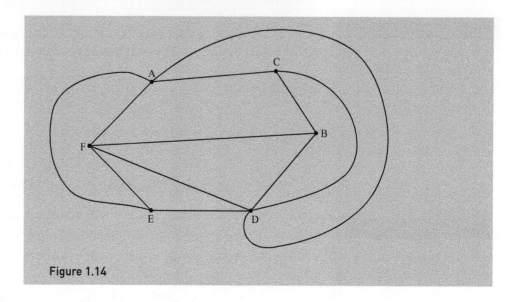

Figure 1.14

Exercise 1.2

① Determine whether the graphs below are

(i)   Eulerian

(ii) semi-Eulerian or

(iii) Hamiltonian.

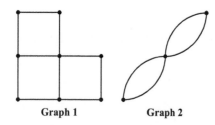

Graph 1   Graph 2

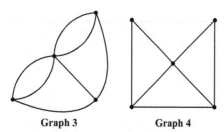

Graph 3   Graph 4

Figure 1.15

② Verify that the graphs in Figure 1.15 satisfy Euler's formula.

③ Show that the graph in Figure 1.16 is planar, and that Euler's formula is satisfied.

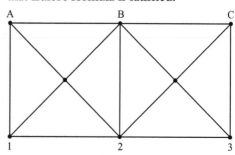

Figure 1.16

④ In the 18th century, the inhabitants of Königsberg (now Kaliningrad) enjoyed promenading across the town's seven bridges – shown in the diagram below. It was known not to be possible to cross each bridge once and once only.

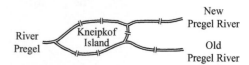

Figure 1.17

(i) Create a graph to model this situation.

(ii) How is it possible to tell that the bridges could not be crossed once and once only?

⑤ The diagram below shows a printed circuit board with two points for external connections and three internal points. Each of the connection points is to be wired to each of the internal points.

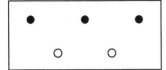

Key

● Internal point

○ Connection point

Figure 1.18

(i) Show that the two connection points can each be wired directly to each of the three internal points without any wires crossing.

(ii) Show that two connection points can be wired to four internal points without any wires crossing.

(iii) Give the smallest numbers of connection points and internal points for which at least one crossing will be required.

[MEI]

⑥ (i) Find a Hamiltonian cycle for this graph.

(ii) Hence show that the graph is planar.

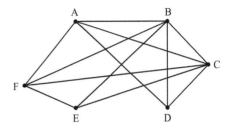

Figure 1.19

⑦ Regarding ABCDEA as different from AEDCBA, how many different Hamiltonian cycles are there in the graph in Figure 1.20?

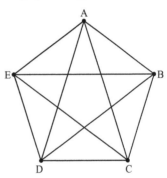

Figure 1.20

3 Further graph theory

Computers are often used to tackle problems involving graphs. The diagram representing a graph is not very convenient for a computer, but it is possible to completely define the essential features of a graph in a way that is usable by a computer.

An **adjacency matrix** (also known as an incidence matrix) shows the number of edges connecting any two vertices. Table 1.3 shows the adjacency matrix for the graph in Figure 1.10.

	A	**B**	**C**	**D**
A	0	1	2	1
B	1	0	1	0
C	2	1	0	1
D	1	0	1	0

Table 1.3

The 2 in row C and column A indicates that there are 2 edges leading from vertex C to vertex A. Notice the symmetry of the matrix about the leading diagonal (from top left to bottom right). This symmetry only applies to a non-directed graph.

Example 1.4

The adjacency matrix for a graph is shown in Table 1.4. Draw the graph it represents.

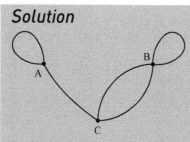

	A	B	C
A	2	0	1
B	0	2	2
C	1	2	0

Table 1.4

Solution

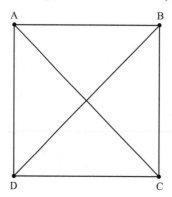

Figure 1.21

The 2 in row A and column A (for example) indicates that there are 2 edges leading from vertex A to vertex A. This counts as a single loop, as it is possible to travel either way round the loop.

A **complete** graph is one where every two vertices share exactly one edge (and where there are no loops). A complete graph with n vertices is denoted by K_n. One representation of the K_4 graph is shown in Figure 1.22.

Figure 1.22

Example 1.5

How many edges does K_n have?

Solution

Labelling the vertices 1 to n, there are $n - 1$ vertices joined by an edge to vertex 1. Excluding vertex 1, there are $n - 2$ vertices joined by an edge to vertex 2, and so on until you reach vertex $n - 1$, which has 1 edge joined to the remaining vertex.

So the total number of edges is $(n - 1) + (n - 2) + \cdots + 1 = \frac{1}{2}(n - 1)n$.

Example 1.6

How many different Hamiltonian cycles does K_n have?

Solution

There will be $\frac{1}{2}(n-1)!$ possible Hamiltonian cycles: you can choose to start at any vertex, and there will be $n-1$ ways of choosing the next vertex to proceed to (and so on). You divide by 2 because reversing the order gives the same cycle.

The **complement** (or inverse) of a simple graph is obtained by adding in the edges necessary to make a complete graph, and then removing the original edges.

Figure 1.23 has been obtained from Figure 1.8 by adding in the necessary edges (shown dotted), in order to obtain K_5. Figure 1.24 has then been obtained by removing the edges contained in the original graph, to give the complement of Figure 1.8.

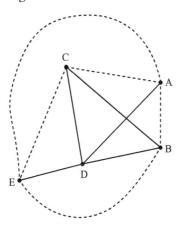

Figure 1.23

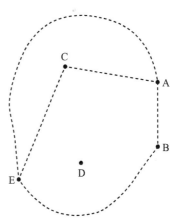

Figure 1.24

Note

Note that whilst a subgraph involves stripping out edges and (isolated) vertices, a subdivision involves adding in vertices and dividing edges into two.

A **subdivision** of a graph is obtained by inserting a new vertex into an edge (one or more times; zero times also counts as a subdivision). Figure 1.25 shows the effect of making two subdivisions of K_4 in Figure 1.22.

Discussion point

→ Is K_4 a subgraph of a subdivision of K_4?

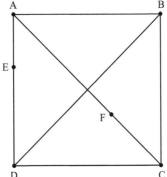

Figure 1.25

Figure 1.26 shows an example of a **bipartite** graph. The special feature of this type of graph is the division of the vertices into two sets, with edges only joining a vertex in one set to a vertex in the other.

A typical application of a bipartite graph is in allocating tasks to workers. Each edge might indicate a task that a particular worker has been trained to perform.

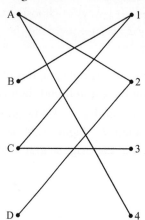

Figure 1.26

In a **complete bipartite graph** (denoted by $K_{m,n}$), each of the m vertices on one side is connected exactly once to each of the n vertices on the other.

Figure 1.27 shows the complete bipartite graph $K_{4,4}$.

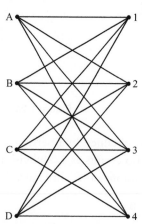

Figure 1.27

Exercise 1.3

① Draw the graph represented by the adjacency matrix in Table 1.5.

	A	B	C	D
A	0	1	2	0
B	1	0	0	1
C	2	0	2	1
D	0	1	1	0

Table 1.5

② Create an adjacency matrix for the graph in Figure 1.28.

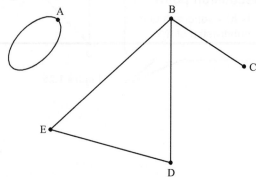

Figure 1.28

③ Write the adjacency matrix for the complement of the graph given by Table 1.6.

	A	B	C	D	E
A	0	0	1	0	1
B	0	0	0	0	0
C	1	0	0	1	1
D	0	0	1	0	1
E	1	0	1	1	0

Table 1.6

④ (i) When is a complete graph Eulerian or semi-Eulerian?

(ii) When is a complete bipartite graph Eulerian or semi-Eulerian?

⑤ Can you find another proof of the fact that K_n has $\frac{1}{2}(n-1)n$ edges?

⑥ Six tasks, A, B, C, D, E and F, are to be carried out by six people, 1, 2, 3, 4, 5 and 6. Each task requires a different combination of people to carry it out. The table shows for which tasks each person is required.

Person	1	2	3	4	5	6
Tasks	A	C	A	D	B	B
	C	D	E	E	E	C
	F	F	F	F	F	F

Table 1.7

Each person can work on only one task at a time.

(i) The situation is to be represented by a graph in which vertices represent tasks. Two vertices are to be joined if the tasks they represent require a person in common. Draw the graph.

(ii) Draw the complement graph and use it to organise the tasks in an efficient way.

(iii) What further information do you need to produce a schedule for the tasks?

[MEI adapted]

⑦ Table 1.8 shows the ski lifts linking four lift stations in a ski resort.

Table 1.9 shows the ski runs linking the stations.

		To		
	Lifts	Maison	Col	Rosa
	Village	✓		
From	Maison		✓	✓
	Col			✓

✓ indicates the existence of a lift
Table 1.8

		To		
	Runs	Village	Maison	Col
	Maison	✓		
From	Col		✓	
	Rosa	✓		✓

✓ indicates the existence of a run
Table 1.9

A skier wishes to ski all of the runs, starting and ending at Village. Which lift must she repeat? You should justify your answer with appropriate working.

[MEI adapted]

⑧ (i) A, B, C and D are the vertices of the complete graph, K_4. List all the paths from A to B.

(ii) Show that there are 16 paths from A to B in the complete graph on the vertices {A, B, C, D, E}.

4 Isomorphisms

Two graphs are said to be **isomorphic** if one can be distorted in some way to produce the other (from the Greek: same form). Isomorphic graphs must have the same number of vertices, each of the same degree, and their vertices must be connected in the same way.

You have already seen how isomorphisms can be used to demonstrate that a graph is planar. The graphs in Figures 1.8 and 1.9 were isomorphic, as they were alternative representations of a planar graph.

Figure 1.8 is repeated as Figure 1.29. The vertices of Figure 1.30 cannot be put into a one-to-one correspondence with those of Figure 1.29 (for example, there is no vertex of degree 4 in Figure 1.30 that could correspond with vertex D of Figure 1.29). Therefore these graphs are not isomorphic.

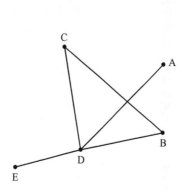

Figure 1.29　　　　　　　**Figure 1.30**

One way of establishing whether two graphs are isomorphic is to compare their adjacency matrices (allowing for the possibility of relabelling the vertices).

Example 1.7

Are the graphs with adjacency matrices shown in Tables 1.10 and 1.11 isomorphic?

	A	B	C
A	2	1	0
B	1	0	2
C	0	2	0

Table 1.10

	A	B	C
A	0	0	2
B	0	2	1
C	2	1	0

Table 1.11

Note

Any two complete graphs with n vertices are isomorphic.

Solution

By comparing the number of edges joining each pair of vertices, it can be seen that the graphs are isomorphic. The vertices have simply been relabelled.

Exercise 1.4

① Determine whether the graphs represented by the following adjacency matrices are isomorphic.

	A	B	C	D
A	0	0	1	2
B	0	0	1	1
C	1	1	2	0
D	2	1	0	2

Table 1.12

	A	B	C	D
A	2	1	0	1
B	1	0	2	0
C	0	2	2	1
D	1	0	1	0

Table 1.13

② Which of the following graphs are isomorphic?

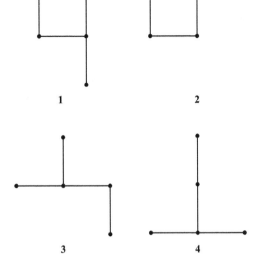

1   2

3   4

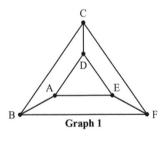

5   6

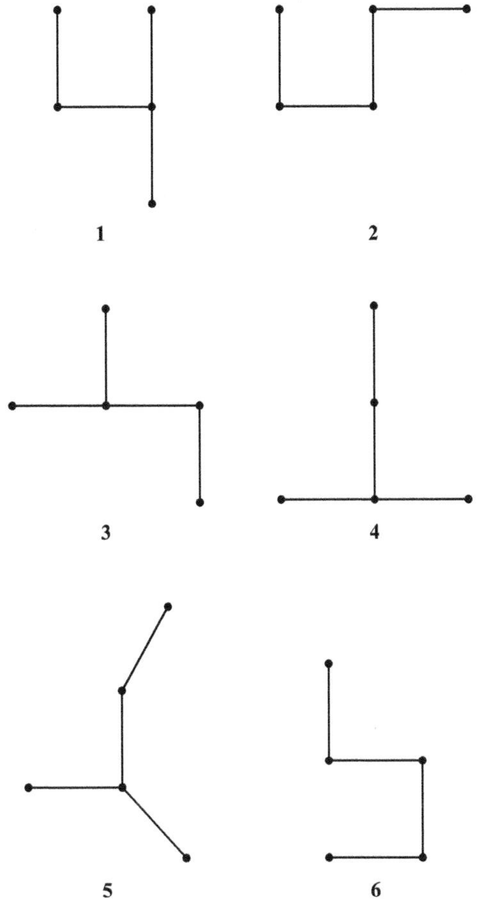

Figure 1.31

③ (i) Show that the following graphs are isomorphic.

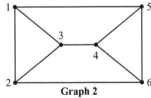

Graph 1

Graph 2

Figure 1.32

(ii) Draw a simple connected graph on six vertices, each of degree 3, which is not isomorphic to Graph 1/Graph 2.

④ Donald claims that the following graphs are isomorphic.

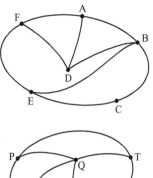

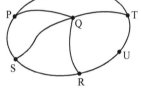

Figure 1.33

Explain why Donald is mistaken.

⑤ Prove that the following graphs are isomorphic.

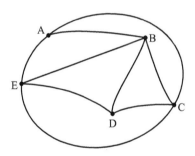

Figure 1.34

	P	Q	R	S	T
P	0	1	1	1	1
Q	1	0	1	1	1
R	1	1	0	1	0
S	1	1	1	0	1
T	1	1	0	1	0

Table 1.14

5 Kuratowski's theorem

Kuratowski's theorem provides a way of determining whether a graph is planar.

The theorem involves the graphs $K_{3,3}$ and K_5 that are shown in Figure 1.35.

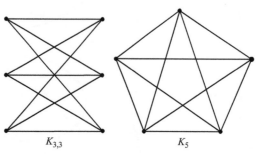

> Remember that a graph is said to be planar if it can be distorted in such a way that its edges do not cross.

$K_{3,3}$ K_5

Figure 1.35

These two graphs can be shown to be non-planar, whilst graphs such as K_4 and $K_{3,2}$, and also $K_{4,2}$, are planar.

Example 1.8

Show that the graphs K_4 and $K_{3,2}$ are planar.

Solution

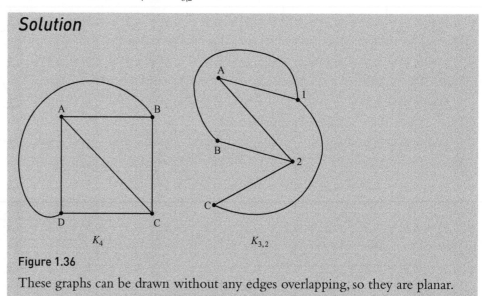

K_4 $K_{3,2}$

Figure 1.36

These graphs can be drawn without any edges overlapping, so they are planar.

A formal statement of Kuratowski's theorem will be given shortly, but for the moment we can say informally that any graph that is simpler than $K_{3,3}$ or K_5 will be planar.

This includes incomplete graphs, as in the following example.

Example 1.9

Show that if an edge is removed from K_5, then the resulting graph is planar.

Solution

Without loss of generality (since the graph is symmetric), you can remove the edge CE.

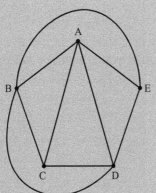

Figure 1.37

There are now no overlapping edges so the graph is planar.

ACTIVITY 1.4

Show that if an edge is removed from $K_{3,3}$, then the resulting graph is planar.

Note

Recall that whilst a subgraph involves stripping out edges and (isolated) vertices, a subdivision involves adding in vertices and dividing edges into two. Also, note that $K_{3,3}$ and K_5 count as subdivisions of themselves.

The formal wording of Kuratowski's theorem is as follows.

A graph is non-planar if and only if it contains a subgraph that is a subdivision of either $K_{3,3}$ or K_5.

So to determine whether a graph is planar, you need to see if it has subgraphs that are subdivisions of either $K_{3,3}$ or K_5.

The examples that follow explain how to do this.

If the graph is bipartite, then you should generally be looking for $K_{3,3}$ as a subgraph; otherwise you will be looking for K_5.

Proof of Kuratowski's theorem

The theorem can be paraphrased informally as

a graph is non-planar if and only if it is as complex as $K_{3,3}$ or K_5.

There are two parts to the paraphrased theorem.

(i)   A graph is non-planar if it is as complex as $K_{3,3}$ or K_5 (or 'as complex as $K_{3,3}$ or $K_5 \Rightarrow$ non-planar'; alternatively, 'planar $\Rightarrow$ not as complex as $K_{3,3}$ or K_5').

(ii) A graph is non-planar only if it is as complex as $K_{3,3}$ or K_5 (or 'non-planar $\Rightarrow$ as complex as $K_{3,3}$ or K_5'; alternatively, 'not as complex as $K_{3,3}$ or $K_5 \Rightarrow$ planar').

Part (i) sounds quite plausible, and is relatively easy to demonstrate. This will be done next. Part (ii) is less obvious, and the proof is beyond the scope of the A Level course.

To prove part (i), you need to show that if a graph G contains a subgraph that is a subdivision of either $K_{3,3}$ or K_5, then G is non-planar.

Suppose that G is planar. Then any subgraph of G will certainly be planar. But then suppose you are told that one such subgraph is a subdivision of either $K_{3,3}$ or K_5. If this subdivision is planar, then either $K_{3,3}$ or K_5 (as appropriate) would have to be planar, which gives a contradiction, as $K_{3,3}$ and K_5 are known to be non-planar. Therefore a graph that is as complex as $K_{3,3}$ or K_5 is non-planar.

Example 1.10

Use Kuratowski's theorem to show that the following graphs are non-planar.

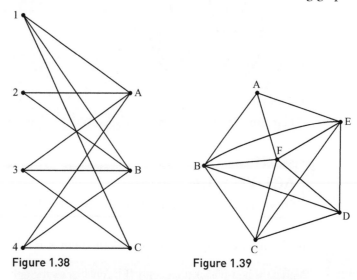

Figure 1.38   **Figure 1.39**

Solution

The bipartite graph in Figure 1.38 has $K_{3,3}$ (with vertices 1, 3, 4, A, B and C) as a subgraph. This is highlighted in Figure 1.40 with darker lines. Therefore, by Kuratowski's theorem, it is non-planar.

Similarly, the graph in Figure 1.39 has K_5 (with vertices B, C, D, E and F) as a subgraph. This is highlighted in Figure 1.41 with darker lines. Therefore it is also non-planar.

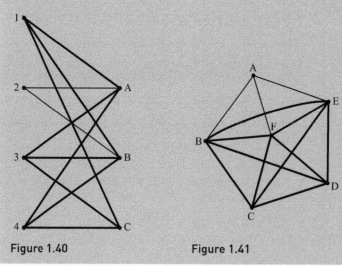

Figure 1.40   **Figure 1.41**

Example 1.11

Use Kuratowski's theorem to show that the following graphs are planar.

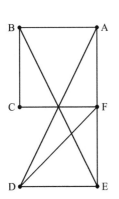

Figure 1.42

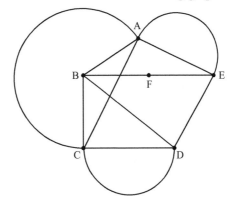

Figure 1.43

Solution

The graph in Figure 1.42 has bipartite graphs as subgraphs, but not $K_{3,3}$. Also, no combination of 5 vertices will produce K_5, so that K_5 isn't a subgraph.

You also need to consider the possibility of there being a subdivision of either $K_{3,3}$ or K_5 within the graph.

Similarly, the graph in Figure 1.43 doesn't contain either $K_{3,3}$ or K_5 (even if you remove F).

Therefore, by Kuratowski's theorem, both graphs are planar.

Exercise 1.5

① Use Kuratowski's theorem to decide whether the following graphs are planar.

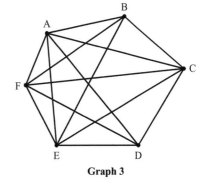

Graph 3

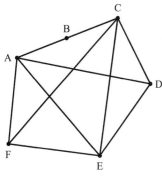

Graph 1

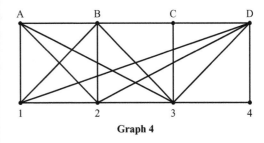

Graph 4

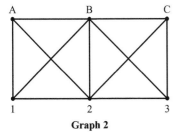

Graph 2

Figure 1.44

② Use Kuratowski's theorem to show that any graph with 8 or fewer edges is planar.

③ (i) Use Kuratowski's theorem to show that $K_{4,2}$ is planar, and confirm this by drawing the graph.

    (ii) Extend this to $K_{n,2}$.

④ Referring to Figure 1.45

    (i)   Is it a subdivision of K_5?

    (ii) Is it a subgraph of K_5?

    (iii) Is it a subgraph of a subdivision of K_5?

    (iv) Does it contain a subgraph that is a subdivision of K_5?

    (v) Is it planar?

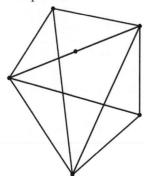

Figure 1.45

⑤ Determine whether the graph below is planar, fully justifying your answer.

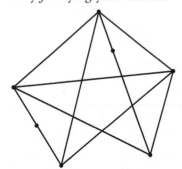

Figure 1.46

⑥ Vertices in a network are said to be adjacent if there is an edge joining them. Colours c1, c2, c3, … are to be assigned to the vertices (a, b, c, …) of a graph so that no two adjacent vertices share the same colour. A method for doing this is as follows:

  1   Allocate colour c1 to vertex a.

  2   Choose the next uncoloured vertex in alphabetical order. List the colours of all the vertices adjacent to it that are already coloured. Choose the first colour that is not in that list and allocate that colour to the vertex.

  3   Repeat step 2 until all vertices are coloured.

  (i) (a) Use the method to complete the colouring of the following network, taking the vertices in alphabetical order.

           Indicate the colours that you allocate (c1, c2, etc.) in the boxes provided.

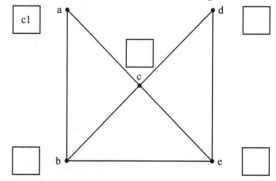

Figure 1.47

    (b) Give a colouring that uses fewer colours.

  (ii) (a) Explain why it is not possible to colour vertices in networks with loops.

    (b) Explain why repeated edges can be ignored when colouring vertices in a network.

  (iii) At a mathematics open day, 6 one-hour lectures a, b, c, d, e and f are to be scheduled so as to allow students to choose from the following combinations.

    a & b   a & d   c & e

    b & f   d & e   e & f   a & f

    (a) Represent this as a network, connecting two vertices with an edge if they must not be scheduled at the same time.

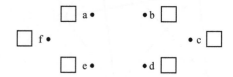

Figure 1.48

    (b) Use the method to colour the network, taking the vertices in alphabetical order. Show the colours that you allocate (c1, c2, etc.) on your network.

    (c) Say how the colouring can be used to give a schedule for the lectures, and give the schedule implied by your colouring.

    (d) Find a best schedule for the lectures.

[MEI adapted]

⑦ There are six people in a room. Prove that there must be at least three people who know each other, or at least three people none of whom know one another.

⑧ There is a number of people at a party, and some handshaking has taken place. Prove that there must be two people who have shaken hands with the same number of people.

LEARNING OUTCOMES

Now you have finished this chapter, you should be able to
➤ understand and use the language of graphs, including vertex, edge, trail, cycle, connected, degree, subgraph, subdivision, multiple edge and loop
➤ identify or prove properties of a graph including that a graph is Eulerian, semi-Eulerian or Hamiltonian
➤ understand and use Euler's formula for connected planar graphs
➤ understand and use complete graphs and bipartite graphs, including adjacency matrices and the complement of a graph
➤ understand and use simple graphs, simple-connected graphs and trees
➤ use Kuratowski's theorem to determine the planarity of graphs
➤ recognise and find isomorphism between graphs.

KEY POINTS

1 A graph consists of vertices and edges.
2 A trail is a sequence of edges in which the end of one edge is the start of the next, and where no edge is repeated.
3 A path is a trail with the further restriction that no vertex is repeated.
4 A cycle is a closed path.
5 A graph is connected if there exists a path between every pair of vertices.
6 A simple graph is one that has no multiple edges or loops.
7 A tree is a simple connected graph with no cycles.
8 The degree of a vertex is the number of edges that join it.
9 A graph is Eulerian if it contains a closed trail that includes all the edges. It is semi-Eulerian if there is a trail that includes all the edges, but isn't closed.
10 A graph is Hamiltonian if it contains a cycle that visits all of the vertices exactly once.
11 A graph is planar if it can be distorted in such a way that its edges do not cross.
12 Euler's formula: $F + V = E + 2$
13 An adjacency matrix shows the number of edges connecting any two vertices.
14 A complete graph is one where every two vertices share exactly one edge (and where there are no loops).
15 The complement of a simple graph is obtained by adding in the edges necessary to make a complete graph, and then removing the original edges.
16 A subdivision of a graph is obtained by inserting a new vertex into an edge.
17 A bipartite graph contains two sets of vertices, with edges only joining a vertex in one set to a vertex in the other.
18 Two graphs are isomorphic if one can be distorted in some way to produce the other.
19 Kuratowski's theorem says that a graph is non-planar if and only if it contains a subgraph that is a subdivision of either $K_{3,3}$ or K_5.

2

Networks

→ What is the shortest route that visits A, B, C and D in the network below and returns to its starting point?

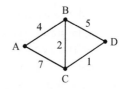

1 The language of networks

When graphs were dealt with in Chapter 1, the terms vertices and edges were used. For networks, it is customary to replace these with **nodes** and **arcs**, respectively.

A network is a **weighted** graph – i.e. a graph for which there is a number (**weight**) associated with each arc.

Figure 2.1 shows a graph from Chapter 1 with weights attached.

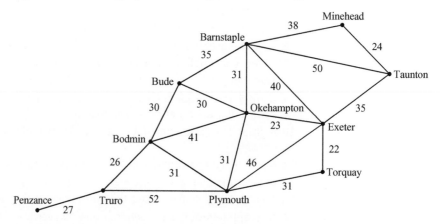

Figure 2.1

In this case, the weights are the distances in miles between the various towns and cities. In other situations the weights could be times or costs, for example.

A **table of weights** (or distance matrix, if applicable) can be used to represent the network in a form that is convenient for a computer. The network in Figure 2.1, which has no loops, multiple arcs or directed arcs, can be represented by Table 2.1.

	Pen	Tru	Ply	Tor	Exe	Tau	Min	Bar	Oke	Bud	Bod
Pen	–	27	–	–	–	–	–	–	–	–	–
Tru	27	–	52	–	–	–	–	–	–	–	26
Ply	–	52	–	31	46	–	–	–	31	–	31
Tor	–	–	31	–	22	–	–	–	–	–	–
Exe	–	–	46	22	–	35	–	40	23	–	–
Tau	–	–	–	–	35	–	24	50	–	–	–
Min	–	–	–	–	–	24	–	38	–	–	–
Bar	–	–	–	–	40	50	38	–	31	35	–
Oke	–	–	31	–	23	–	–	31	–	30	41
Bud	–	–	–	–	–	–	–	35	30	–	30
Bod	–	26	31	–	–	–	–	–	41	30	–

Table 2.1

2 The minimum connector problem

Suppose that a cable company wishes to join up the towns and cities in Figure 2.1, using the shortest possible length of cable. This can be done by creating a spanning tree with the minimum weight.

There are several ways of doing this. One method is simply to remove arcs in order of decreasing weight, ensuring that the network remains connected.

Example 2.1

Find a minimum spanning tree for the network in Figure 2.1.

Solution

The arcs can be removed in the following order (arcs of the same weight may be chosen arbitrarily):

Tru–Ply 52

Bar–Tau 50

Ply–Exe 46

Bod–Oke 41

Bar–Exe 40

Bar–Min 38

Bud–Bar 35

(Exe–Tau 35 can't be removed, as the network would no longer be connected)

Ply–Tor 31

Bod–Ply 31

(Ply–Oke 31 can't be removed)

→

The resulting spanning tree is shown in Figure 2.2. It has a total weight of 279.

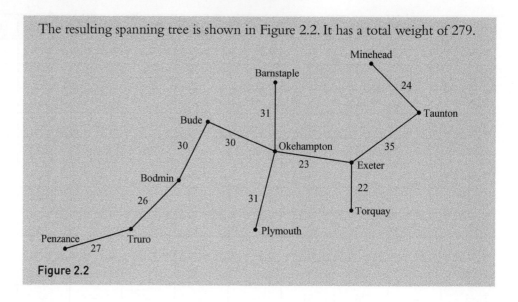

Figure 2.2

Some methods may be easier for a computer to apply. One such method is **Prim's algorithm**.

(i)   Start with any node.

(ii) Add the arc leading to the nearest node.

(iii) Add the arc leading (from any of the nodes collected so far) to the nearest new node, and repeat.

(iv) Stop once all nodes have been collected.

If two nodes are the same distance from the nodes collected so far, either node may be chosen.

Example 2.2

Use Prim's algorithm to find a minimum spanning tree for the network in Figure 2.1, starting from Okehampton. Make clear the order in which the arcs are selected.

Solution

Oke–Exe 23

Exe–Tor 22

Oke–Bud 30

Bud–Bod 30

Bod–Tru 26

Tru–Pen 27

Tor–Ply 31 ◄———— Or Oke–Bar, or Oke–Ply, or Bod–Ply.

Oke–Bar 31

Exe–Tau 35

Tau–Min 24

The resulting spanning tree is shown in Figure 2.3.

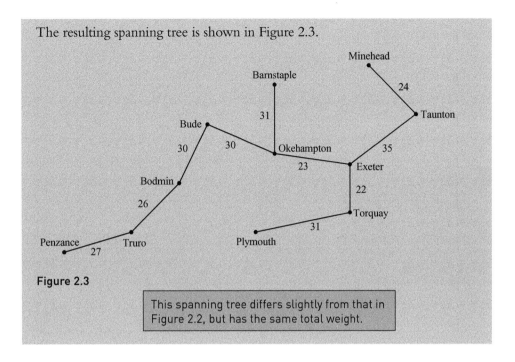

Figure 2.3

> This spanning tree differs slightly from that in Figure 2.2, but has the same total weight.

Another well-known method is **Kruskal's algorithm**:

(i) Start with the shortest arc.

(ii) Choose the next shortest arc, provided it doesn't create a cycle, and repeat. (If two arcs are of equal length, then either may be chosen.)

The order in which arcs are added should be made clear.

Example 2.3

Use Kruskal's algorithm to find a minimum spanning tree for the network in Figure 2.1, starting from Okehampton. Make clear the order in which the arcs are selected.

Solution

Tor–Exe 22

Exe–Oke 23

Tau–Min 24

Bod–Tru 26

Tru–Pen 27

Bod–Bud 30 ← Or Bud–Oke.

Bud–Oke 30

Bod–Ply 31 ← Or Oke–Bar, or Oke–Ply, or Ply–Tor.

Oke–Bar 31 ← Not Ply–Oke or Ply–Tor, as these would create a cycle.

Exe–Tau 35 ← Not Bud–Bar.

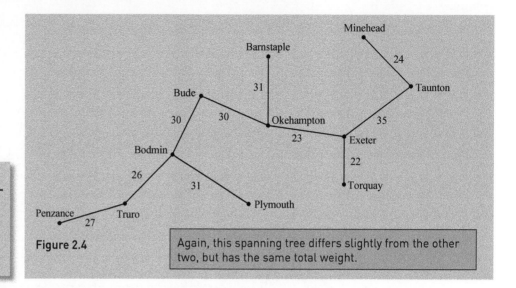

Minehead

Barnstaple

24

31

Taunton

Bude

30   30   Okehampton   35

23   Exeter

Bodmin

26   22

31   Torquay

Penzance   Truro   Plymouth

27

Figure 2.4

Again, this spanning tree differs slightly from the other two, but has the same total weight.

As you have seen, if some of the arcs have the same weight then there may be more than one solution. However, the total weight will always be the same.

Exercise 2.1

① Apply Prim's algorithm to Figure 2.1, starting at Bodmin.

② Create a table of weights to represent the network in Figure 2.5.

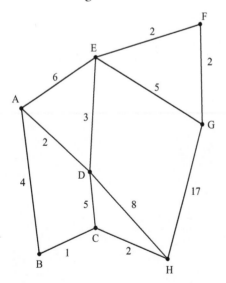

Figure 2.5

③ Draw the network with the table of weights in Table 2.2.

	A	B	C	D	E	F
A	–	10	20	–	16	12
B	10	–	15	–	–	–
C	20	15	–	11	8	–
D	–	–	11	–	7	22
E	16	–	8	7	–	14
F	12	–	–	22	14	–

Table 2.2

④ Apply Prim's algorithm to the network in question 3, starting at A. Show the order in which the steps are carried out, and give the total weight of the minimum spanning tree.

⑤ Apply Kruskal's algorithm to the network in question 3. Show the order in which the steps are carried out.

⑥ Modern Electronic Instruments Ltd. has seven depots. The distances between them, in miles, by direct road links are shown in Figure 2.6.

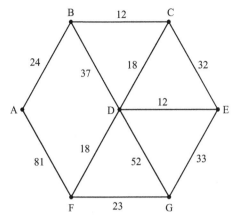

Figure 2.6

Communication cables are to be laid along the roads so that all the depots are linked by cable, either directly or via other depots.

(i) Apply an appropriate algorithm to find the minimum length of cable that is needed, showing the network of cables used.

An eighth depot is established at H. The new depot is connected directly to A, B and E, with distances 6, 17 and 54 miles respectively.

(ii) Find the minimum length of cable needed to link the extended network.

[MEI adapted]

⑦ Five new houses, A, B, C, D and E are to be connected to a drainage system, each having a connection to the sewer at the point S on the diagram. All connections are either direct or via another house. Alternatively, houses may be connected to an intermediate manhole at M. This manhole must in turn be connected to S.

All connecting pipes must be such that water can drain downhill.

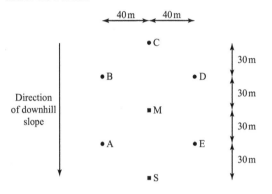

Figure 2.7

(i) Draw a matrix, showing the lengths of all 16 possible connecting pipes.

(ii) Use Prim's algorithm to find a minimum connector for A, B, C, D, E, M and S, giving its total length.

Does your connector represent a system that drains correctly?

(iii) Investigate whether or not the provision of the intermediate manhole at M is worthwhile. Justify your conclusions.

3 The route inspection problem

This is also known as the Chinese postman problem.

In Chapter 1, you met the example of the gritting lorry that needs to pass along all the arcs exactly once. An Eulerian graph was defined as one where this was possible, and in such a way that the lorry would be able to return to its starting point. A semi-Eulerian graph was defined as one where the lorry could pass along all of the arcs exactly once, but would not be able to return to its starting point.

The route inspection problem involves first of all establishing whether the graph (i.e. the network without its weights) is Eulerian, semi-Eulerian or neither. Depending on the type of graph identified, you may then need to make some compromise in order to find the shortest route that covers all of the arcs at least once – returning to the starting point, if required.

You saw in Chapter 1 that the number of odd nodes in a graph has to be even, and that graphs can be divided into the following three categories.

A Those with no odd nodes, which are therefore Eulerian.

B Those with two odd nodes, which are therefore semi-Eulerian.

C Those with four or more odd nodes, which are neither Eulerian nor semi-Eulerian.

In the case of category **A** no compromise has to be made: each of the arcs is covered exactly once, and you end up back at the starting point. The length of the shortest route is just the total of the weights of the network.

Although the existence of such a route is guaranteed, you still need to find it. However, there is usually more than one solution, and it normally isn't difficult to find an example.

In the case of category **B** if it is permissible for the start and end nodes to be different, then no compromise is necessary. However, if you wish to end up where you started, then you have to convert the network into one that has no odd nodes.

This is done by finding the shortest possible path between the two odd nodes, and duplicating that path (so that there are multiple arcs between some pairs of nodes). In this way, the two odd nodes are made even, and any nodes along the duplicated path have their degree increased by two (and so will still be even).

Example 2.4

For the network shown in Figure 2.8, find the shortest route such that all arcs are covered at least once and which returns to the starting point.

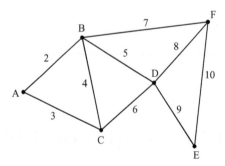

Figure 2.8

> It is not necessary for BF to follow CB in the shortest route: it is sufficient that the arcs CB and BF are repeated at some point.

Solution

Here there are two odd nodes: C and F.

As the shortest route between them is CBF (of weight $4 + 7 = 11$), this path is repeated in the network, as shown in Figure 2.9.

After repeating this path, the network is as in Figure 2.9.

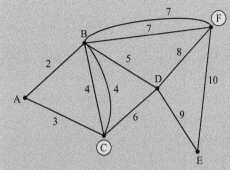

Figure 2.9

The total length of the original arcs is 54, and the effect of adding on the repeated arcs is to give a value of $54 + 4 + 7 = 65$ for the shortest route such that all arcs are covered at least once and which returns to the starting point.

ACTIVITY 2.1

For the network given in Figure 2.9, find a route that starts and ends at A.

In the case of category **C** where there are more than two odd nodes, there are several possibilities.

(i) You must return to the starting point.

(ii) You can start and finish where you like.

(iii) You must start or finish at a specified node.

For **(i)**, the approach is similar to that for category **B**: you decide how to pair up the odd nodes in such a way that the total of the shortest distances between the paired nodes is minimised.

To do this, you first find the shortest path between each possible pair of odd nodes (AB, AC, AD, BC, BD, CD – where the four odd nodes are A, B, C and D).

Then you establish all the possible ways of pairing up the odd nodes (in the case of four nodes, there will be three possibilities: AB & CD, AC & BD and AD & BC).

You then choose the combination of pairings that gives the shortest total path (e.g. AC & BD, if AC + BD is the smallest possible total).

This total is the additional distance that has to be added to the original total of all the arcs, to give the length of the shortest route.

Example 2.5

For the network given in Figure 2.10, find the shortest route that allows all arcs to be covered at least once and which returns to the starting point.

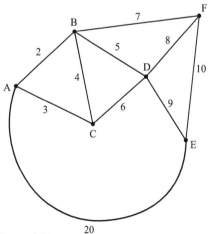

Figure 2.10

ACTIVITY 2.2

For the network given in Figure 2.10, find a route that starts and ends at A.

ACTIVITY 2.3

For Figure 2.10, what would be the best strategy if you had to start at A and finish at B (an example of category C(iii))?

ACTIVITY 2.4

Find the shortest route if you have to start at A and finish at B.

Solution

There are now four odd nodes: A, C, E and F.

The possible pairings, together with the shortest distances associated with them, are (by inspection) as follows.

AC 3

AE 16 (ABDE)

AF 9 (ABF)

CE 15 (CDE)

CF 11 (CBF)

EF 10

The possible ways of pairing up these nodes, together with the total distances in each case, are

(AC) (EF)   3 + 10 = 13

(AE) (CF)   16 + 11 = 27

(AF) (CE)   9 + 15 = 24.

The combination that gives the shortest total distance is thus (AC) (EF).

Since the total of the original arcs is 74, the effect of adding the repeated arcs is to give a value of 74 + 13 = 87 for the shortest route that allows all arcs to be covered at least once and returns to the starting point.

Discussion point

→ In the case of Figure 2.10, what would be the best strategy for category C(ii), where you can choose the start and end nodes?

Example 2.6

Arnold, who is a railway enthusiast, wishes to travel along each stretch of railway linking the cities A–H of a particular country, as shown in Figure 2.11, with the times (in hours) for each stretch. The total time for all the stretches is 57 hours.

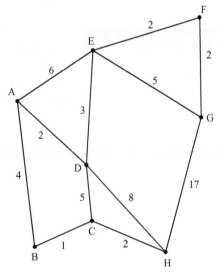

Figure 2.11

(i) Initially he plans to set out from A and return to A. Find a route that covers each stretch of railway at least once, in the shortest possible time, and find the time taken.

(ii) There is a change of plan, and now Arnold wishes to start at A and finish at H (still covering each stretch at least once). Find the new time taken for the quickest route.

(iii) Arnold's wife wants the time reduced. If he has a free choice as to the starting and finishing cities, which should he choose, and what will the new time be?

Solution

(i) First of all, the degrees of each node are established.

A 3 B 2 C 3 D 4 E 4 F 2 G 3 H 3

Thus there are 4 nodes of odd degree: A, C, G and H.

In order to create an Eulerian graph (one where you can return to the starting position, having travelled along each arc exactly once), the nodes of odd degree need to be converted to even degree, by adding in extra arcs (which will be repeats of some of the existing arcs).

For example, you might join up A and C, and then G and H.

The other possibilities are AG and CH, and AH and CG.

In joining up these nodes, you need to use paths between them that have the smallest total weight.

These total weights are as follows.

AC and GH: 5[ABC] + 14[GFEDCH] = 19

AG and CH: 9[ADEFG] + 2[CH] = 11

AH and CG: 7[ABCH] + 12[CDEFG] = 19

So the best option is AG and CH, which involves repeating the arcs AD, DE, EF, FG and CH (as shown in Figure 2.12).

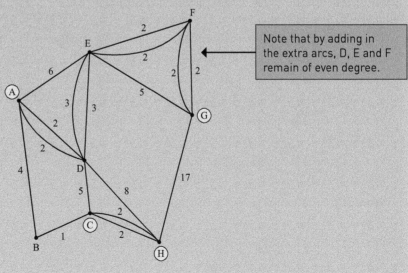

Note that by adding in the extra arcs, D, E and F remain of even degree.

Figure 2.12

One possibility is ABCDADEDHCHGEFGFEA

The time taken is then 57 + 11 = 68 hours.

(ii) Starting at A and finishing at H means that nodes A and H can remain of odd degree, so that only C and G need to be joined up.

This gives rise to an additional 12 hours (adding CDEFG), and hence the total time is now 57 + 12 = 69 hours.

(iii) Of the available options for joining up nodes of odd degree, CH has the smallest weight (of 2). Hence the best option is to start at A and finish at G, or vice versa, so that CH is travelled along twice.

This gives a total time of 57 + 2 = 59 hours.

Exercise 2.2

① (i) Use the route inspection method to find a shortest route that covers all of the arcs in Figure 2.13 at least once, starting and finishing at node 1.

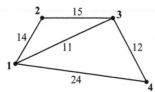

Figure 2.13

(ii) Find a shortest route that covers all of the arcs in Figure 2.13 at least once, starting and finishing at any suitable nodes.

[MEI adapted]

② Find a shortest route that covers all of the arcs in Figure 2.14 at least once, starting at Oxford and returning there. Distances are in miles.

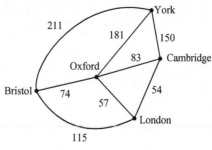

Figure 2.14

③ Find a shortest route that covers all of the arcs in Figure 2.14 at least once, starting at Oxford and ending at a different city.

④ Find a shortest route that covers all of the arcs in Figure 2.14 at least once, starting at one city and ending at a different one.

⑤ Find a shortest route that covers all of the arcs in Figure 2.15 at least once, starting at Okehampton and returning there.

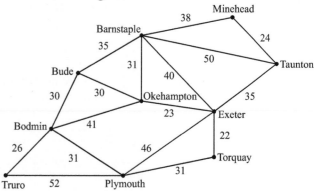

Figure 2.15

⑥ Find a shortest route that covers all of the arcs in Figure 2.15 at least once (you may start and finish at any town or city).

⑦ A highways maintenance depot must inspect all the manhole covers within its area. The road network is given below. In order to make the inspection an engineer must leave the depot, D, drive along each of the roads in the network at least once and return to the depot.

(i) What is the minimum distance that she must drive?

(ii) What route enables her to drive the distance in (i)?

(iii) How many times is node F visited during this route?

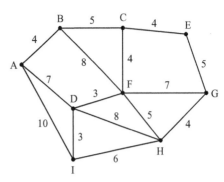

Figure 2.17

4 The travelling salesperson problem

In Chapter 1 a Hamiltonian cycle was defined as one that visits all of the nodes of a network exactly once and which returns to the starting point (another name for a Hamiltonian cycle is a **tour**). In the travelling salesperson problem, the aim is to find the shortest route that visits all of the nodes **at least once**. It may be that this route is a Hamiltonian cycle (i.e. where the nodes are not repeated), but the overriding priority is for the length to be minimised.

Consider, for example, the networks in Figure 2.17 and Figure 2.18.

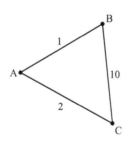

Figure 2.17

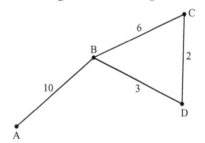

Figure 2.18

> **Note**
> ----------------
> The reason for the significance of the Hamiltonian cycle is that some of the algorithms that can be employed (and which you will be looking at shortly) are designed for Hamiltonian cycles.

In Figure 2.17, a Hamiltonian cycle exists (e.g. ABCA), but you are better off with ABACA, in order to minimise the length.

In Figure 2.18, if A is to be the start and end point, then you cannot avoid repeating B, so that the route will not be a Hamiltonian cycle.

A problem is said to be **classical** if the aim is to find a Hamiltonian cycle (i.e. no node is to be repeated), and it is called **practical** if each node is to be visited **at least** once before returning to the start.

However, a practical problem can always be converted to a classical one by the following device.

■ For each pair of nodes in the network, establish the shortest distance between them (which may be along an indirect path).

■ Then create a complete graph with the given nodes and attach these shortest distances to the appropriate arcs.

For example, the network in Figure 2.18 becomes the one in Figure 2.19, whilst the network in Figure 2.19 becomes the one in Figure 2.20.

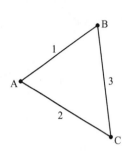

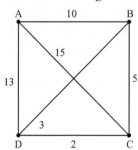

Figure 2.19

Figure 2.20

Unfortunately, there is no one algorithm that will enable you to find a shortest route. However, various methods exist for improving on a route that has already been found (i.e. one which passes through all the nodes, but is not of the shortest possible length).

The length associated with a route that has been found is called an **upper bound**, and so the aim is to reduce the upper bound.

There is also a method for finding a **lower bound** for the shortest distance. Once the upper bound is sufficiently close to the lower bound, you may decide that further effort is not worthwhile.

Finding an initial upper bound

Example 2.7

The network showing the distances in miles between various cities is repeated in Figure 2.21. Find an upper bound for a Hamiltonian cycle.

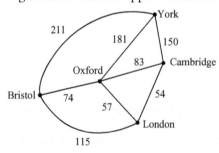

Figure 2.21

Solution

Three possible routes (with their total lengths) are as follows.
BYOCLB (211 + 181 + 83 + 54 + 115 = 644)
BYCOLB (211 + 150 + 83 + 57 + 115 = 616)
BOYCLB (74 + 181 + 150 + 54 + 115 = 574) ◄─── The shortest length cannot be greater than 574, as you have already found a route of this length.
The upper bound (so far) is 574.

Finding a lower bound

You will now look at a method for finding a lower bound (often referred to as the **lower bound algorithm**). This method is only guaranteed to work if the network is complete.

Example 2.8

Use the lower bound algorithm to find a lower bound for a Hamiltonian cycle in the network shown in Figure 2.21.

Solution

Any Hamiltonian cycle will consist of two arcs from (say) B, together with three arcs linking O, L, C and Y.

The shortest possible total length of the two arcs from B is $74 + 115 = 189$.

To find the shortest possible total length of the arcs linking O, L, C and Y, you can find the minimum connector for these nodes.

By removing arcs in decreasing order, the length of the minimum connector for O, L, C and Y is found to be

$(181 + 150 + 83 + 57 + 54) - 181 - 83 = 261$.

Therefore the shortest possible total length of all the arcs (and a lower bound for a Hamiltonian cycle) is $189 + 261 = 450$.

So far, for the network shown in Figure 2.21, it has been established that the shortest possible length of a route that visits all the vertices lies between 450 and 574.

Further lower bounds can be established by dividing up the nodes differently. If node O is isolated instead of B, then the shortest possible total length of the two arcs from O is $57 + 74 = 131$, and using Kruskal's algorithm, for example, for the remaining nodes, you obtain a length of $54 + 115 + 150 = 319$, so that the lower bound is $131 + 319 = 450$ (again).

Isolating the other nodes in turn gives

L + BOCY: $(54 + 57) + (74 + 83 + 150) = 418$

C + BOLY: $(54 + 83) + (57 + 74 + 181) = 449$

Y + BOLC: $(150 + 181) + (54 + 57 + 74) = 516$.

The value of 516 supersedes the other, lower values: although it is true that the shortest route cannot be lower than 418, it is also true that it cannot be lower than 516.

You now have a lower bound of 516 and an upper bound of 574.

Using the nearest neighbour algorithm to find an upper bound

This means that the method gives a possible tour.

The **nearest neighbour algorithm** is a systematic way of finding a solution (and hence an upper bound). The method is only guaranteed to work if the network is complete.

Refer to Figure 2.21 again.

(i)   Start at any node (e.g. B).
(ii) Add the shortest arc leading to a new node: BO.
(iii) Repeat the process, to give BO + OL + LC + CY.

(iv) Return directly to the start node, to give the cycle

BOLCYB $(74 + 57 + 54 + 150 + 211 = 546)$

(v) Repeat the algorithm, with other starting points.

OLCYBO $(57 + 54 + 150 + 211 + 74 = 546)$

LCOBY: can't return to L.

CLOBY: can't return to C.

YCLOBY $(150 + 54 + 57 + 74 + 211 = 546)$

In general, you take the lowest of the values obtained.
You now have a lower bound of 516 and an upper bound of 546.
Therefore the solution is written as

| This is not a solution if it does not give a tour. | → | $516 \leqslant$ optimal solution $\leqslant 546$ | ← | This will be a tour and it may be the optimal solution. |

It is easy to confuse Prim's algorithm with the nearest neighbour algorithm. Make sure you know the difference.

Discussion point

→ What are the differences between the nearest neighbour algorithm and Prim's algorithm?

The nearest neighbour algorithm is a 'greedy' algorithm: it doesn't look ahead, and just maximises the short-term gain, by selecting the nearest node. It doesn't usually give the best possible solution.

Solutions obtained by applying an algorithm can sometimes be improved on.

A possible **tour improvement algorithm** can be illustrated by referring to Figure 2.21 again. The three possibilities that were mentioned earlier were

BYOCLB $(211 + 181 + 83 + 54 + 115 = 644)$

BYCOLB $(211 + 150 + 83 + 57 + 115 = 616)$

BOYCLB $(74 + 181 + 150 + 54 + 115 = 574)$.

Note that BYOCLB is improved by swapping O and C, or by swapping Y and O. The algorithm consists of examining each sequence of 4 nodes, and seeing if an improvement can be obtained by swapping the middle two nodes.

Using computer language, the algorithm could be written as follows (where N_i denotes the ith node, and d denotes distance between nodes).

For $i = 1$ to n
If $d(N_i, N_{i+2}) + d(N_{i+1}, N_{i+3}) < d(N_i, N_{i+1}) + d(N_{i+2}, N_{i+3})$ then swap N_{i+1} and N_{i+2}
Next i

Example 2.9

Apply the lower bound algorithm to the network in Figure 2.22, by isolating Exeter.

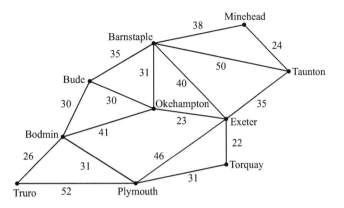

Figure 2.22

Solution

Exclude Exeter and all arcs connected to it.

Then find a minimum connector for the nodes that remain.

You can remove arcs in decreasing order, as follows (ensuring that the network remains connected).

Tru–Ply 52

Bar–Tau 50

Bod–Oke 41

Bud–Bar 35

This leaves the following arcs in the minimum connector.

Tor–Ply 31

Ply–Bod 31

Bod–Tru 26

Bod–Bud 30

Bud–Oke 30

Oke–Bar 31

Bar–Min 38

Min–Tau 24

The total weight is 241.

The two shortest arcs leading from Exeter are 22 and 23: add these to the total weight of the minimum connector.

So the lower bound for the length of the tour is 241 + 22 + 23 = 286.

Note

The lower bound is made as high as possible (i.e. as close to the upper bound as possible) to reduce the interval in which the optimal solution lies.

If the lower bound is equal to the upper bound, then the solution is optimal.

ACTIVITY 2.5

Apply the nearest neighbour algorithm to the network in Figure 2.22, with Taunton as the starting point.

Exercise 2.3

① Apply the lower bound algorithm to the network in Figure 2.11 (repeated in Figure 2.23), isolating each of the nodes in turn. What is the lower bound that it produces?

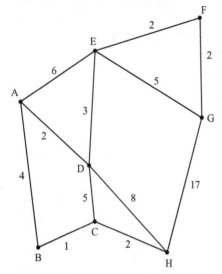

Figure 2.23

② Apply the nearest neighbour algorithm to the network in Figure 2.23, considering each of the nodes as a possible starting point. What is the upper bound that it produces?

③ Apply the lower bound algorithm to the network in Figure 2.24 (from question 3 of Exercise 2.1), isolating each of the nodes in turn. What is the lower bound?

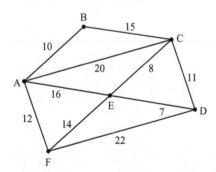

Figure 2.24

④ Apply the nearest neighbour algorithm to the network in Figure 2.24, considering each of the nodes as a possible starting point.

⑤ For the network in Figure 2.25, create the table of weights associated with the complete network obtained by finding the shortest distance between each pair of nodes (i.e. converting a practical problem to a classical problem).

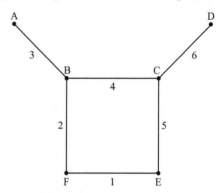

Figure 2.25

⑥ A group of tourists staying in Weston wishes to visit all the places shown on the following map.

 (i) Suggest a route that will minimise their total driving distance.

 (ii) A tree has blocked the road between Weston and Cheddar, making it impassable for the whole day. Suggest an alternative route that will result in the least extra driving distance.

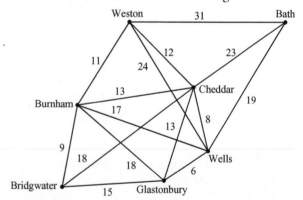

Figure 2.26

⑦ A depot located in Birmingham supplies goods to customers in Sheffield, Nottingham, Stoke, Shrewsbury, Hereford, Gloucester and Northampton.

 (i) Plan a suitable route for the delivery lorry if it has to make deliveries in all of these towns on one trip. The distances involved are shown in the table.

 (ii) Suggest two reasons why the shortest route may not be the best route for the lorry.

	Birmingham	Sheffield	Nottingham	Stoke	Shrewsbury	Hereford	Gloucester	Northampton
Birmingham	–	77	50	43	43	52	52	50
Sheffield	77	–	37	47	79	125	128	94
Nottingham	50	37	–	50	79	102	102	57
Stoke	43	47	50	–	34	83	89	85
Shrewsbury	43	79	79	34	–	52	75	93
Hereford	52	125	102	83	52	–	28	91
Gloucester	52	128	102	89	75	28	–	72
Northampton	50	94	57	85	93	91	72	–

Table 2.3

Exercise 2.4

① A network has four odd vertices, A, B, C and D. One pairing of them is A with B and C with D.

(i) How many ways are there of pairing four odd vertices?

(ii) How many ways are there of pairing six odd vertices?

(iii) How many ways are there of pairing eight odd vertices?

(iv) How many ways are there of pairing twenty odd vertices?

② The graph below is a representation of a system of roads. The lengths of the roads are shown in metres.

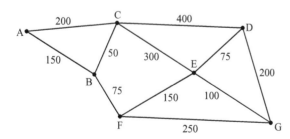

Figure 2.27

(i) List the odd vertices in the graph.

(ii) Explain why the graph is not Eulerian.

(iii) Find the shortest route that starts and finishes at A and traverses each road at least once. State the length of the route.

③ In a sweet-making factory, five flavours of fruit drop are made one after another on a single machine. After each flavour, the machine must be cleaned in readiness for the next flavour. The time spent cleaning depends on the two flavours as indicated in the table below.

Time in minutes	Next flavour to be made				
Last flavour made	Strawberry	Lemon	Orange	Lime	Raspberry
Strawberry	–	14	12	19	16
Lemon	21	–	14	10	19
Orange	19	16	–	17	20
Lime	17	9	13	–	15
Raspberry	20	15	13	19	–

Table 2.4

The production manager wishes to find a sequence that minimises the total time spent cleaning the machine in each cycle from strawberry to strawberry, making each flavour of fruit drop once only per cycle.

(i) By constructing an appropriate network, explain how the problem may be formulated as a travelling salesperson problem. Hence, by using the nearest neighbour algorithm starting from strawberry, suggest a production sequence to the manager.

Someone notices that the smallest cleaning time is in changing from lime to lemon. Accordingly, he suggests that a better production sequence may be found by using the nearest neighbour algorithm starting from lime, so that the sequence will begin with the change from lime to lemon.

(ii) Determine whether he is right that a better sequence will be found.

④ A depot located at town A supplies goods to customers in towns B, C, D and E.

The inter-town distances are given in the table below.

	A	B	C	D	E
A	–	28	57	20	45
B	28	–	47	46	73
C	57	47	–	76	85
D	20	46	76	–	40
E	45	73	85	40	–

Table 2.5

Usually a single vehicle will suffice for a particular delivery but today the customers' requirements are 100 units each and the vehicle available will only carry 300 units. Another similar vehicle can be hired locally, but how should the two vehicles be routed?

(Hint: introduce an artificial depot.)

⑤ (i) When moored alongside a harbour wall a sailing boat has to be secured by four ropes. The ropes are attached to the points A, B, C and D shown in the diagram.

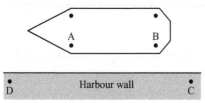

Figure 2.28

The point A has to be attached to D and also to C. The point B has to be attached to C and also to D.

(a) Regard the mooring ropes as defining the arcs of a network and the points A, B, C and D as being the nodes. Give the orders of the nodes.

(b) Explain, in terms of the orders of the nodes, why the network is Eulerian. Indicate the implications of this if only one long length of rope is available instead of four shorter lengths.

(ii) When mooring alongside another boat, ropes are attached both to the other boat and also to the harbour wall, as shown.

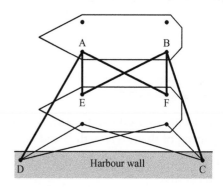

Figure 2.29

Regard the ropes marked with bold lines in the diagram, together with the points A, B, C, D, E and F, as a network.

(a) Give the orders of the nodes and explain why the existence of odd nodes means that the network is not Eulerian.

(b) What practical consequence does the non-Eulerian nature of the network have on the use of a single length of rope to moor the boat?

(c) The solution to the route inspection problem for the network gives a way of mooring the boat with a single length of rope.

Solve the route inspection problem for the network, starting and finishing at the same node, and give the corresponding sequence of nodes.

(d) Suppose that in part (c), the route is allowed to start at C and finish elsewhere. Give two solutions, both starting at C but finishing at different nodes, in which no more than two arcs are repeated.

⑥ The following matrix gives the costs of flight tickets for direct flights between six connected cities.

From \ To	A	B	C	D	E	F
A	–	45	60	58	90	145
B	45	–	67	25	83	100
C	60	70	–	50	70	320
D	50	25	50	–	35	210
E	100	80	70	35	–	72
F	145	110	300	175	80	–

Table 2.6

In this question a tour is a journey from a city, visiting each other city once and only once, and returning to the starting city.

(i) Use the nearest neighbour algorithm to find a tour, starting and finishing at A, with a low associated cost.

Show that the algorithm has not produced the minimum cost tour.

(ii) How many different tours that start and finish at A are there altogether?

(iii) Suppose that, in addition to the cost of tickets, airport taxes must be paid on leaving an airport according to the following table.

A	B	C	D	E	F
20	30	20	40	10	20

Table 2.7

Thus the flight from A to B will cost 45 for the ticket and 20 tax, a total of 65.

Produce a matrix showing total costs, i.e. the total of fares and taxes.

(iv) Give an example where the cheapest route from one city to another differs when taxes are taken into account from that when taxes are not taken into account. Show your two routes and give their costs.

(v) Do airport taxes have an effect on the problem of finding the cheapest tour starting and finishing at A? Why?

[MEI]

⑦ White, yellow, blue, green and red dyes are to be used separately in a dyeing vat (a container in which materials are dyed). Each colour is to be used once during each day. The vat has to be cleaned between colours, and the cost of this depends on which colour was previously used, and on which colour is going to be used. For example, if the blue dye was previously used, and the yellow dye is going to be used next, the cost is 5. These costs, in suitable units, are shown in the table. The vat must be returned to its original colour at the end of the day.

From \ To	W	Y	B	G	R
W	–	0	2	1	2
Y	4	–	4	3	4
B	8	5	–	1	2
G	7	3	1	–	3
R	7	4	3	3	–

Table 2.8

(i) Explain why this problem is similar to the travelling salesperson problem.

(ii) Use the nearest neighbour algorithm five times, starting from each colour in turn, to find a low-cost sequence of colours.

(iii) Give a colour sequence of cost 12, starting and ending with white.

(iv) Because the network is directed, the technique of deleting a node and finding a minimum connector for the remainder to produce a lower bound will not work. Explain why not.

[MEI]

LEARNING OUTCOMES

Now you have finished this chapter, you should be able to

➤ understand and use the language of networks, including: node, arc and weight

➤ solve network optimisation problems using spanning trees

➤ solve route inspection problems

➤ find and interpret upper bounds and lower bounds for the travelling salesperson problem

➤ evaluate, modify and refine models that use networks.

KEY POINTS

1 A network is a weighted graph.

2 The minimum connector problem is solved by creating a spanning tree with the minimum weight. The following methods can be employed.
   ■ Remove arcs in order of decreasing weight.
   ■ Apply Prim's algorithm.
   ■ Apply Kruskal's algorithm.

3 The route inspection problem is to find the shortest route that covers all of the arcs at least once – returning to the starting point, if required. The problem is solved by classifying networks as:
   ■ Eulerian if they have no odd nodes
   ■ semi-Eulerian if they have two odd nodes
   ■ neither Eulerian nor semi-Eulerian if they have four or more odd nodes.

4 The travelling salesperson problem is to find the shortest route that visits all of the nodes of a network at least once, returning to the starting point.

5 A problem is said to be classical if you aim to find a Hamiltonian cycle, and is practical otherwise. A practical problem can always be converted into a classical one.

6 The shortest possible route can be placed within bounds by applying the lower bound algorithm and the nearest neighbour algorithm.

7 Computers may employ a tour improvement algorithm.

3

Linear programming

→ Think about a product that you might make to sell at a fundraising event. What factors affect how much profit you can make?

1 Formulating constrained optimisation problems

Linear programming is a way of solving logistical problems involving linear constraints, as in the following example.

Example 3.1

Millie bakes some cakes for a village fete. She makes cupcakes and chocolate cakes. She sells both types of cake for £1 each.

Each cupcake uses 2 units of sugar and 6 units of flour, and each chocolate cake uses 3 units of sugar and 5 units of flour.

She has only 12 units of sugar and 30 units of flour, and wants to make as much money as possible.

Formulate this as a linear programming problem.

> This is a **constrained optimisation** problem, and the first step is to formulate the constraints mathematically.

Solution

Let x be the number of cupcakes to be made, and y the number of chocolate cakes. The two constraints involving the ingredients become

$2x + 3y \leqslant 12$ (sugar)

$6x + 5y \leqslant 30$ (flour).

Also, $x \geqslant 0, y \geqslant 0.$ ←

> The values of x and y can't be negative.

The total value of sales is $x \times 1 + y \times 1$, and so

$P = x + y$ is the **objective function** that you wish to maximise.

The variables x and y are referred to as the **control variables**.

2 Solving constrained optimisation problems

The cake problem from Example 3.1 can be represented graphically, as shown in Figure 3.1.

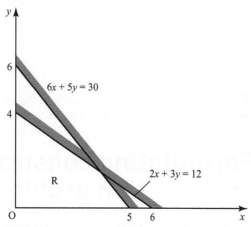

Figure 3.1

The region R in Figure 3.1 is the **feasible region**, where all the constraints are satisfied.

Note that it is the unwanted areas that are shaded. This makes it easier to identify the feasible region – especially where there are a number of constraints.

You now find which points of the feasible region maximise P.

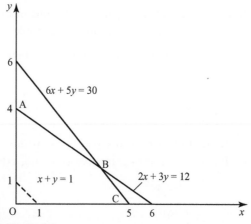

Figure 3.2

> **Note**
> ----
> It can help to use a ruler to represent the moving line, but if it still isn't obvious which vertex is required, you can evaluate the objective function at each vertex, to find the one with the largest value for P.

In Figure 3.2, the line $P = x + y$ is parallel to $x + y = 1$, and needs to be as far away from O as possible, in order to maximise P.

You need to visualise the objective line $x + y = 1$ being moved away from O, keeping its gradient the same, and find the point at which it is about to leave the feasible region. Therefore the maximum value of P will occur at one of the **vertices** (i.e. corners) of the feasible region (unless the line $P = x + y$ has the same gradient as one of the constraint lines). In this example it is B.

Example 3.2

Find the optimal solution for the following linear programming problem.

Maximise $P = x + y$

subject to $2x + 3y \leqslant 12$

$\qquad 6x + 5y \leqslant 30$

$\qquad x \geqslant 0, y \geqslant 0.$

Solution

Look at Figure 3.2.

At B, the lines $6x + 5y = 30$ ① and $2x + 3y = 12$ ② intersect.

Solve these simultaneously.

$3 \times$ ② $-$ ① $\Rightarrow 4y = 6 \Rightarrow y = \frac{3}{2}$

Then ② $\Rightarrow 2x = 12 - \frac{9}{2} \Rightarrow x = 6 - \frac{9}{4} = \frac{15}{4}$

$\Rightarrow P = x + y = \frac{21}{4} = 5.25$ at B (3.75, 1.5).

At A, $P = 4$, and at C, $P = 5$, which confirms that B is the required vertex, so the optimal solution is $x = 3.75$ and $y = 1.5$.

In some contexts, non-integer solutions may be possible.

However, for the problem given in Example 3.1, only integer values are acceptable as x and y represent numbers of cakes. In this case, you need to consider integer points neighbouring (3.75, 1.5), provided they are within the feasible region.

Example 3.3

Solve the linear programming problem given in Example 3.2 if the values of x and y can only take integer values.

Solution

The constraints are that

$6x + 5y \leqslant 30$ and $2x + 3y \leqslant 12$.

Maximise $P = x + y$.

Considering integer points neighbouring (3.75, 1.5) gives

(3, 1): $6x + 5y = 23$ and $2x + 3y = 9$; $P = 4$

(3, 2): $6x + 5y = 28$ and $2x + 3y = 12$; $P = 5$

(4, 1): $6x + 5y = 29$ and $2x + 3y = 11$; $P = 5$

(4, 2): $6x + 5y = 34$ (reject)

> (4, 2) is not in the feasible region.

The points (3, 2) and (4, 1) give equally good solutions. However, that does not guarantee that this is the optimal solution.

> Always state the solution in terms of the context of the problem.

So, in order to maximise the total value of sales, the output of cakes should be either 3 cupcakes and 2 chocolate cakes, or 4 cupcakes and 1 chocolate cake.

Note

(5, 0), vertex C, also gives $P = 5$ so 5 cupcakes is another solution that maximises the objective function, but with only one type of cake!

Instead of maximising sales or profit, you may want to minimise costs, for example. It may be possible to create a feasible region in this case, depending on the constraints.

Example 3.4

Create a feasible region for the following linear programming problem and use the objective line method to find the vertex that gives the optimal solution.

Minimise $P = 2x + y$
subject to $3x + 4y \geqslant 24$
$$y \leqslant 3x$$
$$x \geqslant 0, y \geqslant 3$$

Solution

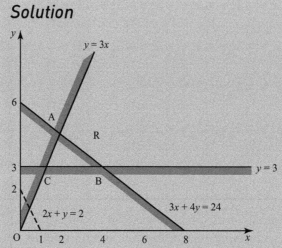

Figure 3.3

Figure 3.3 shows the feasible region, R.

In this case, the line $P = 2x + y$ needs to be parallel to $2x + y = 2$, but as near to the origin as possible.

The diagram shows that P is minimised at A.

ACTIVITY 3.1

Find the optimal solution for the linear programming problem given in Example 3.4, assuming that non-integer solutions are allowed.

Exercise 3.1

① Solve the following linear programming problem.

Maximise $\quad P = x + 2y$
subject to $\quad 4x + 5y \leqslant 45$
$\quad\quad\quad\quad 4x + 11y \leqslant 44$
$\quad\quad\quad\quad x + y \leqslant 6.$

② A farmer grows two crops: wheat and beet. The number of hectares of wheat, x, and the number of hectares of beet, y, must satisfy

$$10x + 3y \leqslant 52$$
$$2x + 3y \leqslant 18$$
$$y \leqslant 4.$$

Determine the values of x and y for which the profit function, $P = 7x + 8y$, is a maximum. State the maximum value of P.

③ A robot can walk at $1.5\,\mathrm{m\,s^{-1}}$ or run at $4\,\mathrm{m\,s^{-1}}$. When walking it consumes power at 1 unit per metre, and when running it consumes power at three times this rate. If its batteries are charged to 9000 units, what is the greatest distance it can cover in half an hour?

④ A builder can build either luxury houses or standard houses on a plot of land. Planning regulations prevent the builder from building more than 30 houses altogether, and he wants to build at least 5 luxury houses and at least 10 standard houses. Each luxury house requires $300\,\mathrm{m^2}$ of land, and each standard house requires $150\,\mathrm{m^2}$ of land. The total area of the plot is $6500\,\mathrm{m^2}$.

Given that the profit on a luxury house is £14 000 and the profit on a standard house is £9000, find how many of each type of house he should build to maximise his profit.

⑤ Maximise $z = x + y$
subject to $3x + 4y \leqslant 12$
$2x + y \leqslant 4$
x integer, y integer.

⑥ The Chief Executive of Leschester City Football Club plc has up to £4 million to spend following a good cup run. He has to decide on spending priorities.

Money needs to be spent on strengthening the playing squad and on extra support facilities (i.e. non-playing staff and stadium facilities).

The Coach, who is popular with the fans, has said that he will resign unless he gets at least £2 million to spend on new players.

The authorities require that at least £0.6 million be spent to remedy stadium deficiencies affecting crowd safety.

Club policy is that the amount to be spent on support facilities must be at least one quarter of the amount to be spent on the playing squad.

(i) Let £x million be the amount to be spent on the playing squad and let £y million be the amount to be spent on support facilities. Write down four inequalities in terms of x and y representing constraints on spending.

(ii) Draw a graph to illustrate your inequalities.

(iii) Find the maximum amount that may be spent on the playing squad.

A report commissioned from a market research company indicates that fans regard both team performance and facilities as being important. The report states that the function $0.8x + 0.2y$ gives a measure of satisfaction with extra expenditure.

The Chief Executive proposes to spend £2.5 million on the playing squad and £1.5 million on support facilities.

(iv) Calculate the measure of satisfaction corresponding to the Chief Executive's proposals.

(v) Add to your graph the line $0.8x + 0.2y = 2.3$, and explain what points on this line represent.

(vi) The Coach argues that the Chief Executive can achieve the same satisfaction score by

spending less in total, but more on the playing squad. How much less and how much more?

[MEI]

⑦ Two products, X and Y, require three ingredients, A, B and C, for their manufacture. Table 3.1 summarises the amounts required and how much of each is available.

		Resource A	Resource B	Resource C
Amount required per unit of product	Product X	15	10	8
	Product Y	5	7	12
Amount available		600	560	768

Table 3.1

It is required to maximise the total output of the two products subject to the amounts available.

(i) Identify variables and formulate an appropriate linear programming problem.

(ii) Solve your linear programming problem graphically, and interpret the solution.

(iii) The amount of B available is increased by 16. Show that the total output can be increased by 1 unit.

(iv) The amount of B available is increased by a further 16. Show that the total output cannot be increased any further.

[MEI]

⑧ Coal arrives at a coal preparation plant from an opencast site and from a deep mine. It is to be mixed to produce a blend for an industrial customer. The customer requires 20 000 tonnes per week of the blend, and will pay £20 per tonne. Deep-mined coal has a marginal cost of £10 per tonne and coal from the opencast site has a marginal cost of £5 per tonne.

The blend must contain no more than 0.17% chlorine, since otherwise the hydrochloric acid produced by burning would corrode the boilers.

The blend must contain no more than 2% sulphur, since this burns to produce sulphur dioxide which subsequently dissolves to give acid rain. Acid rain damages the environment.

The blend must produce no more than 20% ash when burnt, otherwise the boilers will clog.

The blend must contain no more than 10% water, since otherwise the calorific value is affected.

The deep-mined coal has a chlorine content of 0.2%, a sulphur content of 3%, ash of 35% and water 5%. The opencast coal has a chlorine content of 0.1%, sulphur of 1%, ash of 10% and a water content of 12%.

How much of each type of coal should be blended to satisfy the contract with maximum profit? Which constraint is critical and which constraints are redundant?

⑨ A furniture manufacturer produces tables and chairs. A table requires £20 worth of materials and 10 person hours of work. It sells for a profit of £15.

Each chair requires £8 of materials and 6 hours of work. The profit on a chair is £7.

Given that £480 and 300 worker hours are available for the next production batch, find how many tables and chairs should be produced to maximise the profit.

Why might the optimal solution not be a practical solution?

⑩ A clothing retailer stocks two types of jacket which cost her £10 and £30 to purchase. She sells them at £20 and £50 respectively. She needs to order at least 200 jackets and has £2700 to spend.

The cheaper jackets need 20 cm of hanging space. The expensive jackets need only 10 cm each. She has 40 m of hanging space.

(i) Formulate a linear programming problem, assuming that all the jackets will be sold and that the retailer wishes to maximise her profit.

(ii) Solve the problem using a graphical method.

(iii) What would be the effect of trying to increase the order to satisfy a 10% increase in the demand for jackets? Explain your answer.

3 The Simplex algorithm

The Simplex algorithm provides an algebraic method for dealing with linear programming problems, suitable for use by a computer. It can also deal with problems involving more than two variables, which can't be represented graphically.

Example 3.5

Apply the Simplex algorithm to the linear programming problem from Example 3.1.

> This problem involves two variables only.

Solution

First convert the inequalities into equations. $2x + 3y \leqslant 12$ becomes $2x + 3y + s_1 = 12$, where $s_1 \geqslant 0$ is known as a **slack variable**.

The larger the value of s_1, the further you are from the constraint line, and the greater the amount of slack available.

> Slack variables are needed to turn inequalities into equations

Including the objective function, the equations are

$$P - x - y \qquad\qquad = 0 \qquad ①$$
$$2x + 3y + s_1 \qquad = 12 \qquad ②$$
$$6x + 5y \qquad + s_2 = 30 \qquad ③ \qquad\qquad (s_1, s_2 \geqslant 0)$$

The ratio test ensures you are working within the feasible region.

These equations are set up so that all the variables appear on the left-hand side. The equations can also be presented using the **Simplex tableau**, as shown in Table 3.2.

P	x	y	s_1	s_2	Value	Equation
1	−1	−1	0	0	0	①
0	2	3	1	0	12	②
0	6	5	0	1	30	③

Table 3.2

Start with an initial 'solution' of $x = 0, y = 0$. This satisfies the equations above, and gives $P = 0, s_1 = 12, s_2 = 30$.

It corresponds to O in the feasible region in Figure 3.2.

To find a better solution, that gives a larger value for P, you notice that you can (for example) set $x = 0$, so that equations ① – ③ become

$P = y, 3y + s_1 = 12$ and $5y + s_2 = 30$.

This corresponds to working along the y-axis.

You now want to make y as large as possible. Noting the restriction that s_1 and s_2 have to be ≥ 0, it will be possible to set y equal to the lower of $\frac{12}{3}$ and $\frac{30}{5}$; i.e. 4.

This step is referred to as the **ratio test**.

This gives $P = 4, x = 0, y = 4, s_1 = 0$ and $s_2 = 10$, which corresponds to vertex A of the feasible region. This is a feature of the Simplex method, where there are just two control variables. You are working your way round the vertices of the feasible region until no further improvement can be made.

Where the equation for the objective row is $P - 3x - 2y = 5$, for example, it is conventional to aim to maximise x (setting $y = 0$), because this gives the best chance of maximising P, on the grounds that, out of a choice of $5 + 3x$ and $5 + 2y, 5 + 3x$ is more likely to be the larger (though this may not prove to be the case).

Continuing with y as the variable to be maximised, the y column in the Simplex tableau is called the **pivot column**, and the row that provides the maximum value of y from the ratio test is called the **pivot row**. In this example, this is the row corresponding to equation ②.

The pivot row is used to consolidate the improvement in P, as follows.

The aim is to eliminate y from all of the rows except the pivot row. You use a method that a computer can easily apply. You first divide both sides of equation ② by 3, so that the coefficient of y becomes 1. Equation ② is now

$$\frac{2x}{3} + y + \frac{s_1}{3} = 4 \quad ②a$$

Now eliminate y from equation ①, by adding equation ②a to equation ①, to give

$$P - \frac{x}{3} + \frac{s_1}{3} = 4$$

Notice that, by setting x and s_1 equal to zero, you can obtain the improved value of $P = 4$.

→

You also eliminate y from equation ③, by subtracting 5 times equation ②a, to give

$$\frac{8x}{3} - \frac{5s_1}{3} + s_2 = 10.$$

To summarise, the new equations are (relabelling ②a as ⑤)

$$P - \frac{x}{3} + \frac{s_1}{3} = 4 \qquad ④ = ① + ⑤$$

$$\frac{2x}{3} + y + \frac{s_1}{3} = 4 \qquad ⑤ = ② \div 3$$

$$\frac{8x}{3} - \frac{5s_1}{3} + s_2 = 10 \qquad ⑥ = ③ - 5 \times ⑤.$$

The Simplex tableau is now as follows.

P	x	y	s_1	s_2	Value	Equation
1	$-\frac{1}{3}$	0	$\frac{1}{3}$	0	4	④
0	$\frac{2}{3}$	1	$\frac{1}{3}$	0	4	⑤
0	$\frac{8}{3}$	0	$-\frac{5}{3}$	1	10	⑥

Table 3.3

Note that y now appears only once in the pivot column, with a coefficient of 1. Any value chosen for y will not affect the rows in which this variable has a zero coefficient. Similarly, for P and s_2. These variables are termed **basic** variables, whilst the variables that have coefficients in more than one row are termed **non-basic**.

At the end of each stage of the Simplex method, you can obtain the (current) improved solution by setting the non-basic variables to zero. This means that the values of the basic variables will be those in the right-hand column.

So $P = 4, x = 0, y = 4, s_1 = 0$ and $s_2 = 10$.

This corresponds to point A on the graphical representation.

> **Note**
> ----------
> Although the solution has been improved, there is still scope for further improvement, as can be seen from the fact that there is a negative coefficient of x in the (new) objective row of the tableau.

Example 3.6

Apply a second iteration of the Simplex algorithm, to obtain an improved value for P.

Solution

Take x as the pivot column, and apply the ratio test. You see that equation ⑤ gives $\frac{4}{\left(\frac{2}{3}\right)} = 6$, whilst equation ⑥ gives $\frac{10}{\left(\frac{8}{3}\right)} = \frac{15}{4} < 6$.

So equation ⑥ is the pivot row.

> Always show your working for the ratio test and write down the pivot row.

Ensure that the coefficient of x is 1, then obtain new equations by eliminating x from ④ and ⑤, giving

$$P + \frac{s_1}{8} + \frac{s_2}{8} = \frac{21}{4} \qquad ⑦ = ④ + \frac{1}{3} \times ⑨$$

$$y + \frac{3s_1}{4} - \frac{s_2}{4} = \frac{3}{2} \qquad ⑧ = ⑤ - \frac{2}{3} \times ⑨$$

$$x - \frac{5s_1}{8} + \frac{3s_2}{8} = \frac{15}{4} \qquad ⑨ = \frac{3}{8} \times ⑥.$$

The Simplex tableau is now as follows.

P	x	y	s_1	s_2	Value	Equation
1	0	0	$\frac{1}{8}$	$\frac{1}{8}$	$\frac{21}{4}$	⑦
0	0	1	$\frac{3}{4}$	$-\frac{1}{4}$	$\frac{3}{2}$	⑧
0	1	0	$-\frac{5}{8}$	$\frac{3}{8}$	$\frac{15}{4}$	⑨

Table 3.4

Now P, x and y are the basic variables, and the improved solution is

$P = \frac{21}{4}, x = \frac{15}{4}$

> Note that the value for y has reduced from 4 to $\frac{3}{2}$: i.e. the maximum value wasn't needed in the end.

$y = \frac{3}{2}, s_1 = 0$ and $s_2 = 0,$

which corresponds to vertex B of the feasible region.

As the coefficients of s_1 and s_2 in the objective row are both positive, there is no further scope for increasing P.

In the previous example, the aim was to maximise P. If, instead, you wish to minimise P, then this can sometimes be done by maximising $-P$ as in the following example. This also features three control variables (and so couldn't be solved by a graphical method). At this point, a couple of restrictions on the Simplex method should be mentioned. Firstly, inequalities need to be of the form $2x + 3y \leqslant 12$, rather than $2x + 3y \geqslant 12$. This ensures that the initial solution $x = 0, y = 0$ satisfies the constraints. (There are variations of the Simplex method that allow for $\geqslant$ inequalities: the 'two-stage' Simplex and the 'big M' method.) Secondly, when the equations are set up for the constraints, the values on the right-hand side need to be non-negative. This ensures that the ratio test can be carried out in the same way each time. This doesn't apply to the objective row.

Example 3.7

Minimise $2x + y - 3z$, subject to the following constraints.

$x - 4y + z \leqslant 4$

$3x + 2y - z \geqslant -2$

$x \geqslant 0, y \geqslant 0, z \geqslant 0$

Note

Notice that $(0, 0, 0)$ is a feasible solution, which can be taken as the starting point.

Solution

Step 1 Rewrite the problem as

Maximise $P = -2x - y + 3z$

subject to $x - 4y + z \leqslant 4$

and $-3x - 2y + z \leqslant 2$.

> Note that the inequalities are now of the $\leqslant$ type, and that the right-hand values are non-negative.

Step 2 Create equations, with slack variables.

$P + 2x + y - 3z \qquad\qquad = 0 \quad$ ①

$\quad\quad x - 4y + z + s_1 \qquad = 4 \quad$ ②

$\quad -3x - 2y + z \qquad + s_2 = 2 \quad$ ③

Step 3 Represent the equations in a Simplex tableau.

P	x	y	z	s_1	s_2	Value	Equation
1	2	1	−3	0	0	0	①
0	1	−4	1	1	0	4	②
0	−3	−2	1	0	1	2	③

Table 3.5

Step 4 Choose z as the pivot column (as it is the only variable in the objective row with a negative coefficient), and apply the ratio test.

② with $x = y = s_1 = 0 \Rightarrow z = 4$

③ with $x = y = s_2 = 0 \Rightarrow z = 2$

As $2 < 4$, ③ is the pivot row (indicated by circling the coefficient of z in equation 3 in the table below).

P	x	y	z	s_1	s_2	Value	Equation
1	2	1	−3	0	0	0	①
0	1	−4	1	1	0	4	②
0	−3	−2	①	0	1	2	③

Table 3.6

Step 5 Eliminate z from equations 1 and 2.

P	x	y	z	s_1	s_2	Value	Equation
1	−7	−5	0	0	3	6	④ = ① + 3 × ⑥
0	4	−2	0	1	−1	2	⑤ = ② − ⑥
0	−3	−2	1	0	1	2	⑥ = ③

Table 3.7

Step 6 As x now has the largest negative coefficient in the new objective row (equation ④), you choose this as the next pivot column.

You would normally apply the ratio test now, but you note that the coefficient of x in equation ⑥ is negative. As explained in the following note (which you may wish to omit on a first reading), any such row can be excluded when establishing the pivot row.

▌ Note

The ratio test

In step 6 of Example 3.7, you are trying to maximise x, so that when y and s_2 are set equal to zero in equation ④, P (equal to $6 + 7x$) will be increased by as much as possible.

When maximising x you need to ensure that all the constraint equations are satisfied. For equation ⑤, you set s_1 equal to zero (you have already set y and s_2 to zero in equation ④), to leave you with $4x = 2$. This places an upper limit of $\frac{2}{4} = \frac{1}{2}$ on x (x could be less than $\frac{1}{2}$, as it would always be possible to introduce another slack variable, giving $4x + s_3 = 2$).

In equation ⑥, the coefficient of x is negative, and this row cannot be chosen as the pivot row: such a choice would require us to subtract a multiple of the pivot row from the objective row (in order to eliminate x from it), and this would cause the value of the objective row to fall. This assumes that the values in the right-hand column of the constraint rows are non-negative (which is a requirement of the method). So, when applying the ratio test, you only need to consider constraint equations where the coefficient of the pivot variable is positive.

So you take the pivot row to be equation ⑤, and divide by 4 before eliminating x from the other equations, to give equations ⑦–⑨.

P	x	y	z	s_1	s_2	Value	Equation
1	−7	−5	0	0	3	6	④ = ① + 3 × ⑥
0	④	−2	0	1	−1	2	⑤ = ② − ⑥
0	−3	−2	1	0	1	2	⑥ = ③

Table 3.8

P	x	y	z	s_1	s_2	Value	Equation
1	0	$-8\frac{1}{2}$	0	$1\frac{3}{4}$	$1\frac{1}{4}$	$9\frac{1}{2}$	⑦ = ④ + 7 × ⑧
0	1	$-\frac{1}{2}$	0	$\frac{1}{4}$	$-\frac{1}{4}$	$\frac{1}{2}$	⑧ = ⑤ ÷ 4
0	0	$-3\frac{1}{2}$	1	$\frac{3}{4}$	$\frac{1}{4}$	$3\frac{1}{2}$	⑨ = ⑥ + 3 × ⑧

Table 3.9

Step 7 There are no further variables that have negative coefficients in the objective row and positive coefficients elsewhere in their columns, so you have reached the end of the process.

The variables with a single 1 (and otherwise zeros) in their columns are the basic variables, and their values can be read off the right-hand column, whilst the other (non-basic) variables are set to zero.

Solution: $x = 0.5, y = 0, z = 3.5, s_1 = 0, s_2 = 0$.

The maximised value of $-2x - y + 3z$ is 9.5 and so the minimised value of $2x + y - 3z$ is −9.5.

Step 8 Check:

$x - 4y + z = 4$ ($\leqslant 4$) and $3x + 2y - z = -2$ ($\geqslant -2$).

Exercise 3.2

① Use the Simplex algorithm to solve the following linear programming problem.

Maximise $P = 9x + 10y + 6z$

subject to $2x + 3y + 4z \leqslant 3$

$6x + 6y + 2z \leqslant 8$

$x \geqslant 0, y \geqslant 0, z \geqslant 0$.

② Use the Simplex algorithm to solve the following linear programming problem.

Maximise $P = 3w + 2x$

subject to $w + x + y + z \leqslant 150$

$2w + x + 3y + 4z \leqslant 200$

$w \geqslant 0, x \geqslant 0, \ y \geqslant 0, z \geqslant 0$.

③ Use the Simplex algorithm to solve the following linear programming problem.

Maximise $P = 3w + 2x$

subject to $w + x + y + z \leqslant 150$

$2w + x + 3y + 4z \leqslant 200$

$w \geqslant x$ (rewrite this as $x - w \leqslant 0$)

$w \geqslant 0, x \geqslant 0, y \geqslant 0, z \geqslant 0$.

④ The 'Cuddly Friends Company' produces soft toys. For one day's production run it has available $11\,m^2$ of furry material, $24\,m^2$ of woolly material and 30 glass eyes. It has three soft toys that it can produce:

The 'Cuddly Aardvark', each of which requires $0.5\,m^2$ of furry material, $2\,m^2$ of woolly material and two eyes. Each sells at a profit of £3.

The 'Cuddly Bear', each of which requires $1\,m^2$ of furry material, $1.5\,m^2$ of woolly material and two eyes. Each sells at a profit of £5.

The 'Cuddly Cat', each of which requires $1\,m^2$ of furry material, $1\,m^2$ of woolly material and two eyes. Each sells at a profit of £2.

An analyst formulates the following linear programming problem to find the production plan that maximises profit.

Maximise $\quad 3a + 5b + 2c$

subject to $\quad 0.5a + b + c \leqslant 11$

$\qquad\qquad 2a + 1.5b + c \leqslant 24$

$\qquad\qquad 2a + 2b + 2c \leqslant 30$

(i) Explain how this formulation models the problem, and say why the analyst has not simplified the last inequality to $a + b + c \leqslant 15$.

(ii) The final constraint is different from the others in that the resource is integer valued. Explain why that does not impose an additional difficulty for this problem.

(iii) Solve this problem using the Simplex algorithm.

Interpret your solution and say what resources are left over. [MEI adapted]

⑤ A publisher is considering producing three books over the next week: a mathematics book, a novel and a biography. The mathematics book will sell at £10 and costs £4 to produce. The novel will sell at £5 and costs £2 to produce. The biography will sell at £12 and costs £5 to produce. The publisher wants to maximise profit, and is confident that all books will be sold.

There are constraints on production. Each copy of the mathematics book needs 2 minutes of printing time, 1 minute of packing time, and $300\,cm^3$ of temporary storage space.

Each copy of the novel needs 1.5 minutes of printing time, 0.5 minutes of packing time, and $200\,cm^3$ of temporary storage space.

Each copy of the biography needs 2.5 minutes of printing time, 1.5 minutes of packing time, and $400\,cm^3$ of temporary storage space.

There are 10 000 minutes of printing time available on several printing presses, 7500 minutes of packing time, and $2\,m^3$ of tempoary storage space.

(i) Explain how the following initial feasible tableau models this problem.

P	x	y	z	s_1	s_2	s_3	Value
1	−6	−3	−7	0	0	0	0
0	2	1.5	2.5	1	0	0	10 000
0	1	0.5	1.5	0	1	0	7 500
0	300	200	400	0	0	1	2 000 000

Table 3.10

(ii) Use the Simplex algorithm to solve your linear programming problem, and interpret your solution.

(iii) The optimal solution involves producing just one of the three books. By how much would the price of each of the other books have to be increased to make them worth producing? [MEI adapted]

Exercise 3.3

① A factory's output includes three products. To manufacture a tonne of product A, 3 tonnes of water are needed. Product B needs 2 tonnes of water per tonne produced, and product C needs 5 tonnes of water per tonne produced.

Product A uses 5 hours of labour per tonne produced, product B uses 6 hours and product C uses 2 hours.

There are 60 tonnes of water and 50 hours of labour available for tomorrow's production.

(i) Formulate a linear programming problem to maximise production within the given constraints.

(ii) Use the Simplex algorithm to solve your linear programming problem, **pivoting first on the column relating to product C**.

[MEI adapted]

② A farmer has 40 acres of land. Four crops, A, B, C and D are available.

Crop A will return a profit of £50 per acre.
Crop B will return a profit of £40 per acre.
Crop C will return a profit of £40 per acre.
Crop D will return a profit of £30 per acre.
The total number of acres used for crops A and B must not be greater than the total number used for crops C and D.

The farmer formulates this problem as:

Maximise $50a + 40b + 40c + 30d,$

subject to $a + b \leq 20$

$a + b + c + d \leq 40$

(i) Explain what the variables a, b, c and d represent.

Explain how the first inequality was obtained.

Explain why expressing the constraint on the total area of land as an inequality will lead to a solution in which all of the land is used.

(ii) Solve the problem using the Simplex algorithm. [MEI]

③ A production unit makes two types of product, X and Y. Production levels are constrained by the availability of finance, staff and storage space. Requirements for, and daily availabilities of, each of these resources are summarised in Table 3.11.

	Finance (£)	Staff time (hours)	Storage space (m³)
Requirement per tonne for product X	400	8	1
Requirement per tonne for product Y	200	8	3
Resources available per day	2000	48	15

Table 3.11

The profit on these products is £320 per tonne for X and £240 per tonne for Y.

(i) Formulate the problem as a linear programming problem.

(ii) State the objective function, assuming that profit is to be maximised.

(iii) Use a graphical method to find the best daily production plan.

(iv) Set up an initial tableau for the problem and use the Simplex algorithm to solve it. Relate each stage of the tableau to its corresponding point on your graph.

④ In an executive initiative course, participants are asked to travel as far as possible in three hours using a combination of moped, car and lorry. The moped can be carried in the car and the car can be carried on the lorry.

The moped travels at 20 miles per hour (mph) with a petrol consumption of 60 miles per gallon (mpg). The car travels at 40 mph with a petrol consumption of 40 mpg. The lorry travels at 30 mph with a petrol consumption of 20 mpg.

2.5 gallons of petrol are available.

The moped must not be used for more than 55 miles, and a total of no more than 55 miles must be covered using the car and/or lorry.

Find the optimal solution and interpret it in the context of the problem.

⑤ A craft workshop produces three products: xylophones, yodellers and zithers. The times taken to make them and the total time available per week are shown in Table 3.12. Also shown are the costs and the total weekly capital available.

	Xylophones	Yodellers	Zithers	Resource availability
Time (hours)	2	5	3	30
Cost (£00s)	4	1	2	24

Table 3.12

Profits are £180 per xylophone, £90 per yodeller and £110 per zither.

(i) Formulate a linear programming problem to find the weekly production plan that maximises profit within the resource constraints.

(ii) Use the Simplex algorithm to solve the problem, pivoting first on the column of your tableau containing the variable that represents the number of xylophones produced. Explain how your final tableau shows that the workshop should produce 5 xylophones and 4 yodellers.

If, when applying the Simplex algorithm, the first pivot is on the column containing the variable which represents the number of zithers produced, then the final solution obtained is for the workshop to produce 1.5 xylophones and 9 zithers per week.

(iii) How can this production plan be implemented?

(iv) Explain how the Simplex algorithm can lead to different solutions. [MEI adapted]

⑥ A furniture manufacturer is planning a production run. He will be making wardrobes, drawer units and desks. All can be manufactured from the same wood.

He has available 200m² of wood for the production run. Allowing for wastage, a wardrobe requires 5m², a drawer unit requires 3m², and a desk requires 2m².

He has 200 hours available for the production run. A wardrobe requires 4.5 hours. a drawer unit requires 5.2 hours, and a desk requires 3.8 hours.

The completed furniture will have to be stored at the factory for a short while before being shipped. The factory has 50m³ of storage space available. A wardrobe needs 1m³, a drawer unit needs 0.75m³, and a desk needs 0.5m³.

The manufacturer needs to know what he should produce to maximise his income. He sells the wardrobes at £80 each, the drawer units at £65 each and the desks at £50 each.

(i) Formulate the manufacturer's problem as a linear programming problem.

(ii) Use the Simplex algorithm to solve the linear programming problem.

(iii) Interpret the results. [MEI]

⑦ Three chemical products, X, Y and Z, are to be made. Product X will sell at 40p per litre and costs 30p per litre to produce. Product Y will sell at 40p per litre and costs 30p per litre to produce. Product Z will sell at 40p per litre and costs 20p per litre to produce. Three additives are used in each product. Product X uses 5g per litre of additive A, 2g per litre of additive B and 8g per litre of additive C. Product Y uses 2g per litre of additive A, 4g per litre of additive B and 3g per litre of additive C. Product Z uses 10g per litre of additive A, 5g per litre of additive B and 5g per litre of additive C. There are 10kg of additive A available, 12kg of additive B, and 8kg of additive C.

(i) Explain how the initial feasible tableau shown below models this problem.

P	x	y	z	s_1	s_2	s_3	Value
1	−10	−10	−20	0	0	0	0
0	5	2	10	1	0	0	10 000
0	2	4	5	0	1	0	12 000
0	8	3	5	0	0	1	8 000

Table 3.13

(ii) Use the Simplex algorithm to solve your linear programming problem, and interpret your solution.

(iii) The optimal solution involves making two of the three products. By how much would the cost of making the third product have to fall to make it worth producing, assuming that the selling price is not changed?

⑧ Neil is refurbishing a listed building. There are two types of paint that he can use for the inside walls. One costs £1.45 per m² and the other costs £0.95 per m². He must paint the lower half of each wall in the more expensive paint. He has 350m² of wall to paint. He has a budget of £400 for wall paint.

The more expensive paint is easier to use, and so Neil wants to use as much of it as possible.

Initially, the following linear programming problem is constructed to help Neil with his purchasing of paint.

Let x be the number of m² of wall painted with the more expensive paint.

Let y be the number of m² of wall painted with the less expensive paint.

Maximise $P = x + y$

subject to $1.45x + 0.95y \leqslant 400$

$y - x \leqslant 0$

$x \geqslant 0, y \geqslant 0$

(i) Explain the purpose of the inequality $y - x \leqslant 0$.

(ii) The formulation does not include the inequality $x + y \geqslant 350$. State what this constraint models and why it has been omitted from the formulation.

(iii) Use the Simplex algorithm to solve the linear programming problem, pivoting first on the '1' in the y column. Interpret your solution.

The solution shows that Neil needs to buy more paint. He negotiates an increase in his budget to £450.

(iv) Find the solution to the linear programming problem given by changing $1.45x + 0.95y \leqslant 400$ to $1.45x + 0.95y \leqslant 450$, and interpret your solution.

Neil realises that, although he now has a solution, that solution is not the best for his requirements.

(v) Explain why the revised solution is not optimal for Neil. [MEI]

LEARNING OUTCOMES

Now you have finished this chapter, you should be able to

➤ formulate constrained optimisation problems
➤ solve constrained optimisation problems via graphical methods
➤ use the Simplex algorithm for optimising (maximising and minimising) an objective function, including the use of slack variables
➤ interpret a Simplex tableau.

KEY POINTS

1 Linear programming is concerned with solving logistical problems involving linear constraints – constrained optimisation problems.
2 The first step is to formulate the constraints mathematically.
3 The objective function is the item that you wish to maximise (or minimise).
4 The feasible region is the region of the graph where all the constraints are satisfied.
5 The objective function is generally maximised (or minimised) at one of the vertices of the feasible region.
6 Neighbouring points may need to be considered if an integer-valued solution is required.
7 The Simplex algorithm provides an algebraic method for dealing with linear programming problems.

- ■ Inequalities are converted into equations using slack variables.
- ■ A Simplex tableau can be used to standardise the process.
- ■ At each stage of the process, a pivot column is established, and the ratio test is used to identify the pivot row.
- ■ After each stage, the values of any basic variables can be read from the tableau, and the non-basic variables are set to zero.

Critical path analysis

➔ Each morning Paul makes toast using a grill. The grill can take two slices of bread at a time and takes 1 minute to toast each side of the bread. How long does it take Paul to toast three slices of bread?

1 Constructing a network using activity-on-node

Critical path analysis uses networks to help schedule projects involving a number of activities, some of which require other activities to take place before they can begin.

The starting point is the **precedence table** (also known as a dependence table), as shown in Table 4.1.

Activity	Immediately preceding activities	Duration (in hours)
A	–	4
B	–	5
C	A, B	6
D	B	2
E	C	4
F	C, D	3

Table 4.1

From this you can construct a network that places the activities in their correct relation to each other; i.e. taking account of the precedences.

There are two conventions that can be adopted for the network. The one that you will employ in this chapter is called **activity-on-node**. The alternative convention is called activity-on-arc.

In this chapter, you shall be using the terms nodes and arcs, rather than vertices and edges.

Example 4.1

Draw the network for the precedence table shown in Table 4.1.

Solution

The network is shown in Figure 4.1.

All of the activities feed into an end activity (of duration 0), either directly or indirectly.

The information for each activity is recorded at the nodes, with the durations being placed in the middle section of the bottom half of each box (the other two sections will be filled in later).

Note

The end activity may not always be included. It is helpful when there is no final activity in the project.

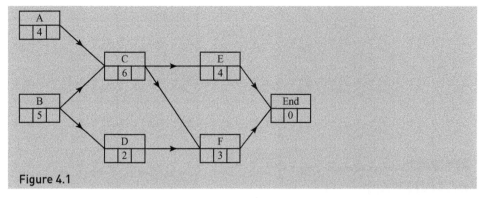

Figure 4.1

Whereas the route inspection and travelling salesperson problems involve an individual (for example) travelling through the network, so that they can only be in one place at a time, critical path analysis typically involves a number of people working simultaneously on different activities, so that different parts of the network are in operation at the same time.

2 Determining earliest start times and latest finish times

You are usually concerned with how soon the whole process can be completed. This is determined by making a **forward pass** through the network, recording the **earliest start time** for each activity.

Look at the network in Figure 4.1 above.

- Activity C cannot start until 5 hours have elapsed, in order that both the activities A and B have been completed. So 5 is entered in the lower left-hand section in C's box (and for completeness a 0 is entered there for A and B).
- 5 is entered for D
- 11 is entered for F, as C will not be completed until 6 hours after its earliest start time of 5 (whilst D can be completed by time 7).

Figure 4.2 shows the situation once all the earliest start times have been determined. The whole process can be completed in 15 hours. This is the **minimum completion time** (or **critical time**).

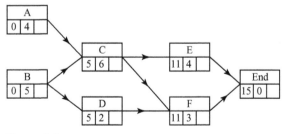

Figure 4.2

Some of the activities may have some slack available. For example, A could be delayed by up to 1 hour without increasing the duration of the project. Those activities that don't have any slack available are called **critical activities**.

In order to establish the extent of any available slack, you carry out a **backward pass**, starting at the end node.

- Place a 15 in the right-hand section for the end activity, to indicate its **latest finish time**.
- As the end activity has duration 0, E and F will also need to finish by time 15, in order for the project not to overrun.
- To find the latest finish time for C you note that it leads into both E and F. You first subtract the duration of E from its latest finish time, to give $15 - 4 = 11$, and then do the same for F, to give $15 - 3 = 12$. So C need only finish by time 12 on account of F, but has to finish by time 11 on account of E, and so its latest finish time is 11.
- For D, which only leads into F, the latest finish time is $15 - 3 = 12$.

ACTIVITY 4.1

Suppose that an error is discovered in the precedence table for Example 4.1, and that activity E should be dependent on activity D (instead of C). Draw a revised activity-on-node network, and find the new minimum completion time.

Figure 4.3 shows the situation once all the latest finish times have been determined.

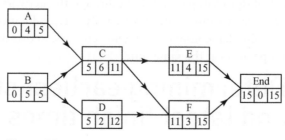

Figure 4.3

Exercise 4.1

① The table shows the precedences for the four tasks of a project. The duration of each task is also shown.

Task	Immediate predecessors	Duration (days)
A	–	2
B	–	1
C	A	1
D	A, B	3

Table 4.2

Draw an activity-on-node network for the project, showing the earliest start and latest finish times. Give the minimum completion time.

② Draw the activity-on-node network for the project with the activities listed below.

Show the earliest start and latest finish times.

Activity	Immediate predecessors	Duration (days)
A	–	8
B	–	4
C	A	2
D	A	10
E	B	5
F	C, E	3

Table 4.3

③ The table shows the activities involved in building a short length of road to bypass a village. The table gives their durations and their immediate predecessors.

	Activity	Immediate predecessors	Duration (weeks)
A	Survey sites	–	8
B	Purchase land	A	22
C	Supply materials	–	10
D	Supply machinery	–	4
E	Excavate cuttings	B, D	9
F	Build bridges and embankments	B, C, D	11
G	Lay drains	E, F	9
H	Lay hardcore	G	5
I	Lay bitumen	H	3
J	Install road furniture	E, F	10

Table 4.4

Draw an activity-on-node network for these activities, showing the earliest start and latest finish times. Give the minimum completion time.

④ A construction project involves nine activities. Their immediate predecessors and durations are listed in the table.

Activity	Immediate predecessors	Duration (days)
A	–	5
B	–	3
C	–	6
D	A	2
E	A, B	3
F	C, D	5
G	C	1
H	E	2
I	E, G	4

Table 4.5

Draw an activity-on-node network for the project, showing the earliest start and latest finish times. Give the minimum completion time.

⑤ The tasks involved in setting an examination paper, their immediate predecessors and durations (in weeks), are given in the following precedence table.

Task		Immediate predecessors	Duration (weeks)
A	Examiner sets questions and produces solutions	–	4
B	Printers produce first draft	A	6
C	Revisers produce recommendations	A	4
D	Examiner produces revisions to the paper	C	1
E	Examiner corrects printers' first draft, incorporating revisions	B, D	1
F	Printers produce second draft	E	4
G	Independent assessor works paper	E	5
H	Revisers examine assessor's report and printers' second draft	F, G	2
I	Examiner makes any changes that are required following assessor's report	G	1
J	Examiner considers revisers' comments and prepares final corrections for printers	H, I	1

Table 4.6

Draw an activity-on-node network for the project, showing the earliest start and latest finish times. Give the minimum completion time.

⑥ Draw a network and carry out the analysis for the following activities involved in building a house. What is the minimum completion time and which activities are critical?

	Activity	Preceding activities	Duration (days)
A	Prepare foundations	–	7
B	Order bricks	–	5
C	Order tiles	–	12
D	Lay drains	A	7
E	Erect shell	A, B	10
F	Roofing	C, E	4
G	Flooring	F	5
H	Plumbing	D, G	12
I	Glazing	G	1
J	Wiring	G	10
K	Plastering	H, J	6
L	Fittings	I, K	2
M	Clear site	I, K	2
N	Paint and clean	L	6
O	Lay paths	M	2

Table 4.7

⑦ The stages involved in recording and promoting a CD are shown in the table below.

	Activity	Preceding activities	Duration (weeks)
A	Record the performance	–	10
B	Design the cover	–	9
C	Book adverts in press	–	3
D	Recording to CD	A	2
E	Produce cover	B	4
F	Packing	D, E	1
G	Promotion copies to radio, etc.	D, E	1
H	Dispatch to shops	F	3
I	Played on radio	G	2
J	Adverts in press, etc.	C, H, I	1

Table 4.8

Draw a network and carry out the analysis to find how long the project will take and which activities are critical.

3 Identifying critical paths

The **float** of an activity is the available slack.

Float = (latest finish time − earliest start time) − duration

Critical activities have zero float.

Example 4.2

Establish the critical activities and floats for the network shown in Figure 4.3 on page 60.

Note

The critical activities B, C and E form a path through the network, leading from the start to the end. Such a **critical path** will always exist, though there may be more than one.

The length of the critical path is the minimum completion time of the project. If there is more than one such path, they will all have the same length.

Solution

A: $5 - 0 - 4 = 1$; float of 1

B: $5 - 0 - 5 = 0$; critical activity

C: $11 - 5 - 6 = 0$; critical activity

D: $12 - 5 - 2 = 5$; float of 5

E: $15 - 11 - 4 = 0$; critical activity

F: $15 - 11 - 3 = 1$; float of 1

ACTIVITY 4.2

Given that all of the activities in the network shown in Figure 4.3 overrun by 1 hour, find the new minimum completion time for the project.

It may be that the completion time of the project has to be reduced, and that this can be achieved by using extra workers to reduce the durations of the critical activities. This is sometimes referred to as **crashing the network**. Any such reductions may change the activities that make up the critical path.

Example 4.3

Table 4.9 shows the costs of reducing the durations of each of the activities in Example 4.1 by one hour, as well as the minimum duration that is possible for each activity.

Activity	Original duration (in hours)	Cost of reducing duration by 1 hour	Minimum duration possible
A	4	£100	2
B	5	£200	4
C	6	£100	3
D	2	£300	1
E	4	£200	1
F	3	£200	2

Table 4.9

Suppose that it is necessary to complete the project as quickly as possible, and you wish to know the extra cost involved in doing this.

Create the new activity network, based on the minimum durations, and establish the critical activities and floats.

Solution

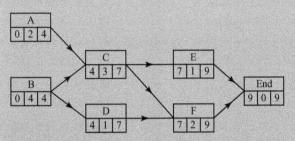

Figure 4.4

One way of tackling this problem is to repeat the original process, using the minimum durations, and then to increase the durations of any non-critical activities.

The new minimum completion time is 9 hours.

A: $4 - 0 - 2 = 2$; float of 2

B: $4 - 0 - 4 = 0$; critical activity

C: $7 - 4 - 3 = 0$; critical activity

D: $7 - 4 - 1 = 2$; float of 2

E: $9 - 7 - 1 = 1$; float of 1

F: $9 - 7 - 2 = 0$; critical activity

You can now increase the durations of the non-critical activities, in order not to incur unnecessary costs. The results are shown in Table 4.10. Note that, for activity D, there is no need to increase the duration above the original figure.

Activity	Original duration (in hours)	Cost of reducing duration by 1 hour	Minimum duration possible	New duration	Extra cost
A	4	£100	2	4	£0
B	5	£200	4	4	£200
C	6	£100	3	3	£300
D	2	£300	1	2	£0
E	4	£200	1	2	£400
F	3	£200	2	2	£200

Table 4.10

The total extra cost is £1100.

① For the project in question 2 of Exercise 4.1, the duration of any of the activities can be accelerated, at a cost.

Table 4.11 shows the normal duration and cost for each activity, the minimum time to which its duration can be reduced and the cost for this time. If an activity is accelerated to a duration greater than the minimum, the cost is calculated on a pro-rata basis. For example, the accelerated cost of activity A to 7 days would be £150 (normal cost $+ \frac{1}{2} \times$ extra cost).

Activity	Duration	Normal cost (£)	Duration	Accelerated cost (£)
A	8	100	6	200
B	4	150	2	350
C	2	50	1	90
D	10	100	5	400
E	5	100	1	200
F	3	80	1	100

Table 4.11

Assuming that the only possible way of reducing the total time is by increasing the costs, which activity or activities would you recommend should be accelerated if the reduction in total time necessary were

(i) 2 days

(ii) 7 days?

② (i) Give the critical activities for the project in question 3 of Exercise 4.1.

(ii) Each of the tasks E, F, G and J can be speeded up at extra cost. The maximum number of weeks by which each task can be shortened, and the extra cost per week saved, are shown in the table below.

Task	E	F	G	J
Maximum number of weeks by which task may be shortened	3	3	1	3
Cost per week of shortening task (in thousands of pounds)	30	15	6	20

Table 4.12

(a) Find the new shortest time for the bypass to be completed.

(b) List the activities which will need to be speeded up to achieve the shortest time found in part (a), and the times by which each must be shortened.

(c) Find the total extra cost needed to achieve the new shortest time. [MEI]

③ (i) Give the critical activities for the project in question 4 of Exercise 4.1.

(ii) Some activity durations can be reduced at extra cost, as shown in Table 4.13.

Activity	Amount by which duration can be reduced	Cost
A	1 or 2 days	£1000 per day
B	1 or 2 days	£500 per day
C	1 or 2 days	£600 per day
D	1 day	£250
E	1 day	£250

Table 4.13

Show that the cheapest way of reducing the minimum project duration by one day involves reducing the durations of two activities. Give the activities and the cost. Show how to reduce the minimum project duration by a further day, and give the cost.

④ (i) Table 4.14 shows activities involved in a project and their immediate predecessors. It is not possible to predict exactly how long each will take, but shortest and longest possible times are shown.

Activity	Predecessors	Shortest possible time (weeks)	Longest possible time (weeks)
A	–	9	19
B	–	10	11
C	B	8	10
D	B	7	7
E	A	5	5
F	A, D	6	21
G	C, E	12	13

Table 4.14

(a) Showing your working, find the shortest possible duration in which the entire project can be completed. List the critical activities.

(b) Showing your working, find the duration for the entire project if all activities take their longest possible times. List the critical activities.

(c) It is suggested that for each activity the shortest and longest activity times be replaced by the mean of the two, so that an estimate can be produced of the actual duration of the entire project. This leads to a project duration of 32 weeks, which is *not* equal to the mean of the shortest and longest durations in (a) and (b). Explain why not.

(ii) In the original project specification activity G could not begin until activities C and E had both been completed. A change in specification allows G to proceed when *either* C or E is completed.

(a) Working with the mean duration times, which are listed below, find the duration of the entire project under these new circumstances.

Activity	A	B	C	D	E	F	G
Mean of short and long times	14	10.5	9	7	5	13.5	12.5

Table 4.15

(b) If the entire project is to be completed in the duration found in part (ii)(a), what is the earliest time at which activity C can be started? What is the latest time by which it must be finished? [MEI]

⑤ A boat purchased in kit form is to be assembled and launched as quickly as possible. The following activities have to be performed, according to the given precedences.

	Activity	Preceding activities	Duration (days)
A	Sand hull	–	2
B	Fit doors	–	0.5
C	Cut out and fit windows	–	1
D	Fit sea cocks and install toilet	A, B	1
E	Fit galley	B	0.25
F	Paint hull and deck	D	3
G	Apply antifouling	F	1
H	Fit winches	C	0.5
I	Step mast	H	0.25
J	Fit internal linings	C	2
K	Fit upholstery	D, E, J	0.5
L	Launch	G, I, K	0.5

Table 4.16

(i) Draw an activity network for the project and determine the minimum completion time.

(ii) The owner does not wish to start fitting the upholstery until the deck paint is dry, i.e. until 4 days after ending activity F. Show how to modify the network to account for this. Find the new critical path and the minimum completion time to launching.

⑥ Claire wants to prepare and eat her breakfast in the minimum time. The activities involved, their immediate predecessors and their durations are shown in Table 4.17.

Activity		Immediate predecessors	Duration (mins)
F	Fill kettle	–	0.5
I	Put instant coffee in cup	–	0.5
W	Boil water	F	3
G	Grill toast	–	7
D	Dish out cereal	–	0.5
O	Fetch and open milk	–	0.5
M	Make coffee	I, W	0.5
B	Butter toast	G	0.5
E	Eat cereal (with milk)	D, O	3
T	Eat toast	E, B	5
C	Drink coffee	M, T	3

Table 4.17

(i) Draw an activity-on-node network for these activities. Do not take account of the fact that Claire can do only one thing at a time.

(ii) Show on your network the earliest start time and the latest finish time for each activity.

(iii) Give the critical activities and the minimum time needed for Claire to complete her breakfast, again taking no account of the fact that she can do only one thing at a time.

(iv) Activities W and G do not require Claire's attention. For all the other activities Claire can do only one thing at a time.

Produce a schedule for Claire starting at 7 am.

At what time does she actually finish her breakfast? [MEI adapted]

⑦ On the last day of a holiday two friends want to spend as much time as possible on the beach before getting ready to catch the plane home. When they leave the beach they will need to clean up and pack, and they want to prepare and eat a meal before they go. The table gives a list of the activities which need to be completed, and their precedences.

Activity		Immediate predecessors	Duration (minutes)
A	Walk from beach to apartment	-	10
B	Dry swimwear	A	60
C	Shower	A	5 (each)
D	Prepare meal	A	20
E	Eat meal	C, D	20
F	Clear up meal	E	20
G	Pack	B, E	15
H	Move cases to reception	F, G	5

Table 4.18

The two showers must be taken at different times.

(i) (a) Produce an activity network representing these activities. Represent taking showers as two consecutive five-minute activities, C1 and C2.

(b) Mark the earliest start time and the latest finish time for each activity in boxes on your activity network. Give the minimum duration and list the critical activities.

(ii) The friends can save time by not waiting for their swimwear to dry before they start packing – they can throw it into the cases at the last moment.

(a) Show how to change the activity network to model this. (It will be sufficient for you to re-draw only that part of the network which needs to be changed.)

(b) Give the reduced duration.

⑧ Table 4.19 shows the durations and precedences for the six activities of a project.

Activity	Duration (days)	Immediate predecessors
A	2	-
B	3	-
C	4	A, B
D	2	B
E	4	C
F	2	C, D

Table 4.19

(i) (a) Draw an activity network for the project.

(b) Mark the earliest start time and the latest finish time for each activity in boxes on your activity network. Give the minimum duration and list the critical activities.

(ii) An extra activity, X, is to be incorporated in the project. The immediate predecessors for X are A and B, and it is to be completed before activity F can begin.

Give the maximum duration allowable for X without increasing the total duration of the project.

(iii) Activity X is redesigned so that it can begin as soon as activity A is completed, and so that it is no longer dependent on activity B.

(a) Produce a revised activity network for the project, incorporating the redesigned activity X.

(b) Find the maximum duration allowable for the redesigned X without increasing the total duration of the project.

(iv) Redesigning activity X so that it is completely independent of activity B is found not to be possible. However, it can be started when activity B is two-thirds complete (and when activity A is completed). Show how to incorporate this information in an activity network. You need only show that part of the network relevant to the starting of activity X.

[MEI]

4 Gantt charts and resource histograms

In addition to establishing the critical activities of an operation, it will usually be important to take account of the resources (people and equipment) required at different stages, and to try to make use of them in the most efficient way.

Draw a **Gantt** (or cascade) **chart** to display the activities shown in the network in Figure 4.3 on page 60.

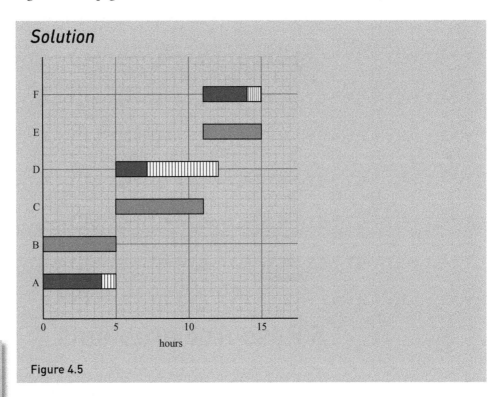

Figure 4.5

Gantt charts can be constructed in slightly different ways, but generally have the following features.

■ Activities are represented by horizontal bars, beginning at their earliest start times.

■ Each activity should be drawn on a separate line. However, critical activities can be drawn on a single horizontal line.

■ The bars are divided into two parts: the length of the first part is the duration of the activity, whilst the length of the second part is its float (the second part can be indicated by shading or a dotted border).

■ The number of workers required for an activity may be shown on the bar.

One use of the Gantt chart is to establish which activities definitely have to be taking place at a given point in time; i.e. activities that can't be shifted away from a particular vertical line.

A **resource histogram** can be drawn, based on the Gantt chart. It shows the number of workers required at any given time.

Example 4.5

Suppose that the activities of the network shown in Figure 4.3 on page 60 require the following numbers of workers.

A: 2   B: 3   C: 1   D: 5   E: 2   F: 4

Draw the resource histogram based on the Gantt chart shown in Figure 4.5.

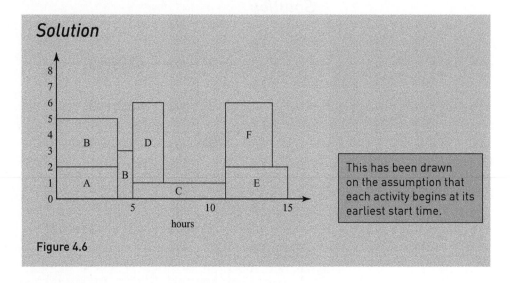

Solution

This has been drawn on the assumption that each activity begins at its earliest start time.

Figure 4.6

5 Resource problems

A factory is likely to employ a workforce on a full-time basis, and will want to keep its workers occupied, as far as possible. It wants to avoid taking on expensive temporary workers. For these reasons, it is desirable to smooth out the number of workers required over the duration of a project. This is referred to as **resource levelling**. It is an example of a **heuristic procedure** (when the solution is likely to be reasonably good, but is not guaranteed to be optimal).

The first step is to examine the resource histogram, to see if the floats allow activities to be shuffled around, in order to limit the number of workers required, without exceeding the critical time.

In the case of Figure 4.6 (relating to Example 4.1), you can see that there is no scope for reducing this limit below 6, without extending the duration of the project. In other words, a workforce of 6 is needed, if temporary workers are not to be employed.

The lowest possible value for this limit can be established by considering areas in the resource histogram. The best you can hope for is that the resource levelling produces a rectangle, with an area equal to the sum of the areas of the blocks in the original resource histogram. This sum is

$$\Sigma \text{ (duration of activity} \times \text{number of workers required for the activity)}$$

where Σ denotes summation over all the activities.

The height of the (ideal) rectangle is then obtained by dividing the area by the base, which is the critical time.

Example 4.6   Find the lowest possible value for the workforce needed in the case of Table 4.20.

Solution

Activity	Duration (in hours)	Workers needed	Duration × number of workers
A	4	2	8
B	5	3	15
C	6	1	6
D	2	5	10
E	4	2	8
F	3	4	12
			Total = 59

Table 4.20

This is the same project that was looked at in Example 4.1. The critical time for this project was 15 hours.

The area of the ideal rectangle is $8 + 15 + 6 + 10 + 8 + 12 = 59$.

Dividing by the critical time gives $59 \div 15 = 3.93$

so a workforce of 4 is the smallest possible number needed.

> However, as you have seen, the actual workforce needed is 6, and in any case, activity D alone requires 5 workers.

Exercise 4.3

① The tasks involved in decorating a room are given in the table.

	Task	Immediate predecessors
A	Strip old paper	–
B	Rub down wooden surfaces	–
C	Paint ceiling	A
D	Apply undercoat	A, B
E	Apply gloss paint	D
F	Paper walls	C, E

Table 4.21

(i) Draw an activity-on-node network to illustrate this information. The duration of each task is shown in Table 4.22.

Task	A	B	C	D	E	F
Duration (days)	1	0.25	0.75	1	1	1

Table 4.22

(ii) Complete forward and backward passes to find the earliest start and latest finish times. Give the critical activities and the minimum duration of the project.

(iii) Produce a resource histogram, given that each task requires one person. Fred is decorating the room on his own. How long will it take him?

(iv) Fred asks Alice to help him. Show that together they could decorate three rooms within eight days. [MEI]

② The table shows activities involved in a construction project, their durations, and their immediate predecessors.

	Activity	Immediate predecessors	Duration (weeks)
A	Obtain planning permission	–	6
B	Survey site	–	2
C	Dig foundations	A, B	6
D	Lay drains	A, B	3
E	Access work	A	10
F	Plumbing	C, D	2
G	Framework	C	6
H	Internal work	E, F, G	4
I	Brickwork	G	3

Table 4.23

(i) Draw an activity network for the project.

(ii) Perform a forward pass and a backward pass to find earliest start and latest finish times. Give the critical path and the minimum time to completion.

(iii) The contractor winning the contract has only one JCB (a digging machine) available. This is needed for activities C (digging foundations) and D (laying drains). Decide whether or not the contractor can complete the project within the minimum time found in (ii). Give reasons and working to support your conclusion. [MEI]

③ (i) The activities involved in clearing and covering an outside court at a major tennis tournament are shown in Table 4.24. Also shown are their immediate predecessors and the minimum time required to complete them.

	Activity	Immediate predecessors	Duration (mins)
A	Escort players and officials off court	–	1
B	Disconnect electronic sensors	–	0.5
C	Remove umpire's chair	A, B	1
D	Remove 1 line judge's chair	A, B	0.5
E	Remove net	B	0.5
F	Remove the two net posts	E	1
G	Pull out waterproof cover	C, D, F	0.5

Table 4.24

(a) Draw an activity-on-node network for these activities.

(b) Mark on your diagram the earliest start times and the latest finish times for the activities.
Give the minimum completion time and the critical activities.

(c) Draw a Gantt chart, given that all activities are scheduled to start as early as possible.

(ii) The numbers of staff required for activities A, B, C and E are 1, 1, 2 and 2, respectively. It takes one person 0.5 minutes to remove one line judge's chair. It takes one person 1 minute to remove one net post. To pull out the waterproof cover takes eight people 0.5 minutes, or it takes four people 1 minute.

(a) Give the minimum number of staff needed to complete the clearing and covering of the court in the minimum total time. Which activity has to be resourced at less than the maximum level to achieve this?

(b) Give the minimum time in which the clearing and covering can be completed if only four staff are available. [MEI]

④ The table shows the activities involved in a project, their durations, and their immediate predecessors.

Activity	A	B	C	D	E	F	G	H
Duration (days)	1	2	1	1	4	2	3	2
Immediate predecessors	–	–	A	B	B	C, D	E, F	E

Table 4.25

(i) Draw an activity-on-node network for the project.

(ii) Perform a forward pass and a backward pass on your network to determine the earliest start and latest finish times. State the minimum time for completion and the activities forming the critical path.

(iii) Draw a Gantt chart for the project, given that all the activities are scheduled to start as early as possible.

(iv) The number of people needed for each activity is as follows.

Activity	A	B	C	D	E	F	G	H
People	1	4	2	3	1	2	3	2

Table 4.26

Activities C and F are to be scheduled to start later than their earliest start times so that only five people are needed at any one time, whilst the project is still completed in the minimum time. Specify the scheduled start times for activities C and F. [MEI]

⑤ The table shows the activities involved in a project, their durations, their immediate predecessors and the number of people needed for each activity.

Activity	A	B	C	D	E	F
Duration (days)	1	2	1	1	4	2
Immediate predecessors	–	–	A, B	B	B	C, D
People needed	1	3	2	2	2	2

Table 4.27

(i) Find the minimum completion time for the project.

(ii) Schedule activities C, D and F so that the project is completed in the minimum time, but so that no more than four people are needed at any one time.

⑥ A company manufactures widgets. The manufacture of a widget involves ten tasks, as shown in the table below.

Task	A	B	C	D	E	F	G	H	I	J
Immediate predecessors	–	–	–	A	A, B, C	C	C	E, F, D	E, F	G

Table 4.28

(i) Construct an activity-on-node network for this project.

Each task normally requires one worker, and durations in hours are given in the table below.

Task	A	B	C	D	E	F	G	H	I	J
Duration (hours)	2	3	4	4	4	3	3	4	2	3

Table 4.29

(ii) Perform forward and backward passes to determine the earliest start times and latest finish times.
Give the critical activities and the minimum duration.

(iii) Show that three workers have to be available to produce a widget in this minimum time, and show how this can be achieved.

The duration of task D can be reduced by 1 hour by employing one extra worker on that task, or by 2 hours by employing two extra workers.

Similarly, the duration of task E can be reduced by 1 or 2 hours by employing one or two extra workers on that task.

(iv) Give the reduced manufacturing time that is achievable, and say which duration(s) should be reduced to achieve this.

[MEI]

⑦ The table below gives information about a construction project.

Activity	Immediate predecessors	Duration (days)
A	–	10
B	–	3
C	B	5
D	A, C	3
E	B	4
F	D	7
G	C, E	2

Table 4.30

(i) Construct an activity-on-node network for this project.

(ii) Perform a forward pass and a backward pass on your precedence network to determine the earliest start times and latest finish times.

(iii) State the minimum time for completion and the critical activities.

(iv) Use an appropriate method to produce an ordering of the activities, and hence draw a Gantt chart for the project.

[MEI]

⑧ The network shows the precedences and durations, in days, of the activities in a project.

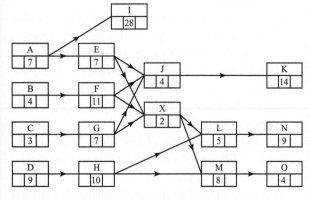

Figure 4.7

(i) Find the earliest start times and the latest finish times of the activities in the network.

(ii) The network in Figure 4.8 is part of a larger network. It shows all of the activities that are connected to activity X, *and only those activities*, together with some earliest start times and latest finish times. Say which of the following it is possible to work out from the information given in this network. Find the value if it is possible to do so, and explain why if it is not.

(a) Earliest start time for U
(b) Earliest start time for X
(c) Latest finish time for Q
(d) Latest finish time for X

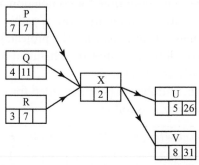

Figure 4.8

⑨ The activity network below shows the durations (in days) of the nine activities of a project and the activities that precede them.

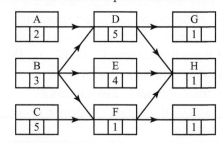

Figure 4.9

(i) Produce a table showing, for each activity, the immediately preceding activities.

(ii) Find the earliest start time and latest finish time for each activity. List the critical activities and give the minimum completion time.

(iii) Draw a Gantt chart, given that all activities are scheduled to start as early as possible. The activities require people as shown in the table below.

Activity	A	B	C	D	E	F	G	H	I
People	1	1	2	1	4	1	1	1	1

Table 4.31

(iv) Draw a resource histogram for the project, indicating clearly which activities are taking place at any given time.

(v) If only six people are available find the shortest possible time in which the project can be completed. State which activities need to be delayed to achieve this.

[MEI adapted]

⑩ A project involves six tasks as follows.

Task	Immediate predecessors	Duration (mins)
A	-	15
B	-	20
C	A, B	10
D	A	6
E	B, D	25
F	C	23

Table 4.32

(i) Draw an activity network for the project.

(ii) Find the minimum completion time and the critical path.

(iii) Give the float times for each non-critical task.

An extra condition is now imposed – task A may not begin until task B has been underway for at least 6 minutes.

(iv) Draw a new network to take account of this. What difference does it make to the minimum completion time and to the critical path?

LEARNING OUTCOMES

Now you have finished this chapter, you should be able to

- construct, represent and interpret a precedence (activity) network using activity-on-node
- determine earliest start and latest finish times for an activity network
- identify critical activities, critical paths and the floats of non-critical activities
- refine models and understand the implications of possible changes in the context of critical path analysis
- construct and interpret Gantt (cascade) charts and resource histograms
- carry out resource levelling (using heuristic procedures) and evaluate problems where resources are restricted.

KEY POINTS

1   Critical path analysis uses networks to help schedule projects involving a number of activities.
2   Based on a precedence table, an activity-on-node network is constructed.
3   Forward and backward passes are made to determine earliest start and latest finish times.
4   A critical path is established, in order to find the minimum completion time.
5   Extra resources may be employed to reduce the durations of the critical activities.
6   Gantt charts and resource histograms can be used to display the activities.
7   Resource levelling may be employed to limit the number of workers used.

5

'Oh, could I flow like thee, and make thy stream my great example, as it is my theme!'

Sir John Denham,
Cooper's Hill

Network flows

→ List all the possible paths from A to D in the network below.

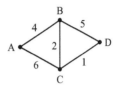

1 The language of network flows

A common problem that may need to be solved is finding the maximum flow that is possible through a network. For example, you might consider the hourly rate of traffic passing from one point to another, or the rate of flow of water through a system of pipes.

Figure 5.1 shows a directed network. In the context of network flows, the weights on the arcs are called **capacities**.

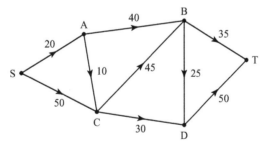

Figure 5.1

However, it is important to note that these weights could represent one of two things: either the actual flow along an arc, or the maximum flow allowed along an arc (i.e. its maximum capacity). In this chapter the terms 'capacity' and 'upper capacity' will both mean the maximum flow allowed along an arc (and you will see shortly that an arc may have a 'lower capacity'). Any actual flows will be described as such (rather than as 'capacities').

Figure 5.1 shows the upper capacities, and the task is to establish the actual flows needed along all the arcs, so that the flow through the whole network is maximised. Note that when the word flow is used, it means a flow in a given time period; i.e. a rate of flow.

 Note

It is not permissible for there to be a build-up at any of the nodes, and so the total flow out of each node must equal the total flow into that node.

The starting node is called the **source** (usually denoted by S) and the end node is called the **sink** (usually denoted by T). Sources are nodes that have all directed arcs pointing outwards, whilst sinks are nodes that have all directed arcs pointing inwards. There may be more than one source or sink.

Example 5.1

Determine the maximum flow across the network given in Figure 5.1.

Solution

Start by noting that a flow of 30 is possible, by just using the path SCDT (i.e. with no flow along the other arcs). This gives the network in Figure 5.2.

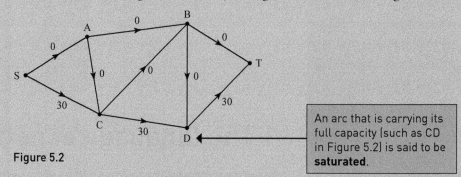

An arc that is carrying its full capacity (such as CD in Figure 5.2) is said to be **saturated**.

Figure 5.2

This flow of 30 can be increased by using other arcs.

For example, SABT provides an additional flow of 20, to give a total of 50.

You can improve this by using SCBT (this can support a flow of 35), together with SABDT (which can support a flow of 20), to give a total of 55.

A further improvement can then be made by also using SCDT (noting that you have already used a flow of 35 along SC, as well as a flow of 20 along DT), and this gives us a further flow of 15, and hence a total of 70.

To summarise, the following flows are possible.

SCBT: 35

SABDT: 20

SCDT: 15

These flows are combined in Figure 5.3.

> **Note**
>
> As a check, note that the total flow out of each node equals the total flow into that node (except for the source and sink).

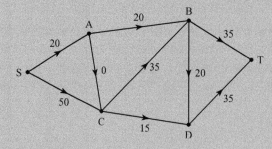

Figure 5.3

2 Maximum flow–minimum cut theorem

A useful device for maximising the flow is the **cut**. In general, a cut is any line or curve that divides the network into two parts, separating S and T. It must not pass through a node.

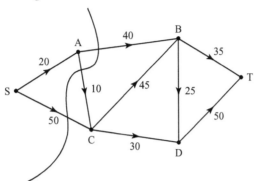

Figure 5.4

An example is shown in Figure 5.4. Imagine that you are dealing with a network of roads in neighbouring countries, and that the cut is the border between the countries.

A cut can be specified in the form: {source-side nodes}/{sink-side nodes}.

The cut in Figure 5.4 would be written as {S, A}/{B, C, D, T}.

The **value** of the cut (also known as the capacity of the cut) in Figure 5.4 is 40 + 10 + 50 = 100. ◄──── Note that the flows that you are using for the cut are the maximum allowed flows along the various arcs.

This is the maximum flow across the border (in a given time period), arising from the three crossing points.

This is a simple case, where the three flows are all crossing the border in the same direction (i.e. from the left-hand country to the right-hand country).

Figure 5.5 shows a cut where one of the arcs (DB) has a flow from the right-hand country to the left-hand country. As you are trying to maximise the flow from left to right, there is no need to have any flow from B to D (as this will reduce the total flow), and so the value of the cut is defined as the sum of the flows from left to right only; i.e. 40 + 45 + 50 = 135.

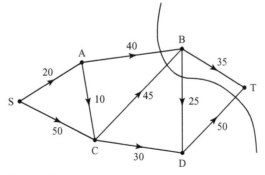

Figure 5.5

Considering the example of the two countries again (A and B, say), you can interpret S and T as their capital cities. Suppose that the location of the border has yet to be decided. The possible cuts in the network are where the border could be placed. The number of tourists, for example, that can pass from S to T in a given time period cannot exceed the number that pass from country A to country B across any particular border, represented by a cut.

In general, the maximum possible flow across a network can be determined by finding the cut that has the smallest value. This is the **maximum flow–minimum cut theorem**.

> **Note**
> ----------------
> In Example 5.1, a flow of 70 was found by inspection. As the cut {S}/{A, B, C, D, T} has a value of 70, you know that this flow of 70 cannot be improved upon.

> The value of the maximum flow = the value of the minimum cut.

However, for large networks it may be time-consuming to establish all the possible cuts. Although there is no algorithm associated with this particular theorem that enables the maximum flow to be found, the theorem does mean that if you can find a flow and a cut of the same value, then you will have found the maximum flow (as it is not possible for the flow to exceed the value of the cut). However, you shall see that an algorithm does exist for finding the maximum flow.

Exercise 5.1

①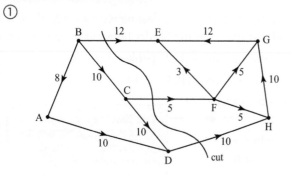

Figure 5.6

(i) In the network shown above, identify the source and the sink.

(ii) Find the capacity of the indicated cut.

② (i) Find minimum cuts for each of these networks.

(a)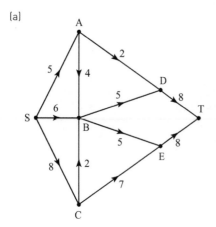

Figure 5.7

(b)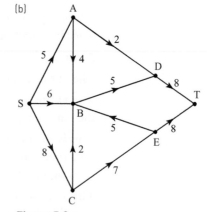

Figure 5.8

(ii) Confirm that your cuts are minimum cuts by finding a flow of the same value in each network.

③ List all the cuts in the network shown in Figure 5.9, and find their values. Hence determine the maximum flow through the network, and find a flow pattern that has this maximum value.

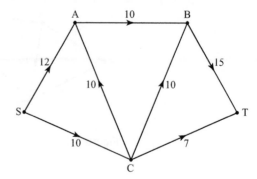

Figure 5.9

④ The following diagram shows capacities in a network.

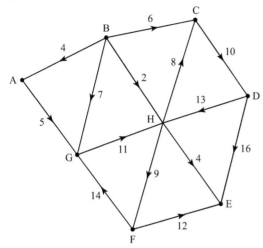

Figure 5.10

(i) Identify the source and sink.

(ii) Find the values of the cuts {A,B,G}/{C,D,E,F,H} and {B,C,H}/{A,D,E,F,G}

(iii) What do these values tell us about the maximum flow through the network?

⑤ The following diagram shows actual flows in a network. Determine a, b, c, d and e.

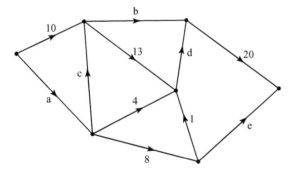

Figure 5.11

⑥ The following diagram shows capacities in a network.

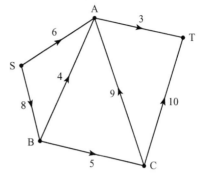

Figure 5.12

(i) Identify all the possible cuts and give their values.

(ii) Hence determine the value of the maximum flow.

(iii) Find a flow that has this maximum value.

3 Multiple sources and sinks

Figure 5.13 shows an example where there is more than one source, and more than one sink. This situation can easily be dealt with by introducing a **supersource** and a **supersink**. This is shown in Figure 5.14. Large capacities (relative to those for the existing arcs) are introduced for the new arcs created, so that they don't affect the solution. Alternatively, the capacity of SS_1 could just be set to 40 (10 + 30), as this is the maximum flow that could leave S_1. Also the capacity of T_2T could just be set to 50 (20 + 30), and similarly for the other sources and sinks.

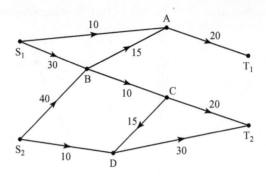

Figure 5.13

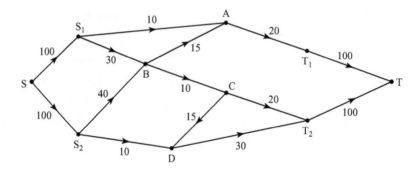

Figure 5.14

4 Flow augmenting

The algorithm referred to earlier for finding the maximum flow involves a process known as **flow augmenting** (or the labelling procedure).

Using our earlier example, you start with any flow that you can find, such as the one in Figure 5.2, with a flow of 30 along SCDT. For each arc of the network, you note how much scope there is for increasing and decreasing the flow. These two figures (i.e. the potential increase and the potential decrease) are then indicated on a new version of the network, by replacing each arc with two arcs having opposite directions, as shown in Figure 5.15. (It is also possible to keep the single arcs, but attach suitable labels to them.)

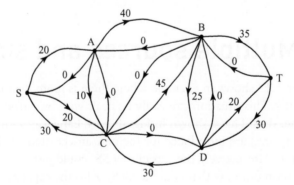

Figure 5.15

You then look for any flow within this network of potential changes. For example, the flow of 20 along SABT (as found previously). You note this change, and modify the network of potential changes, to give Figure 5.16.

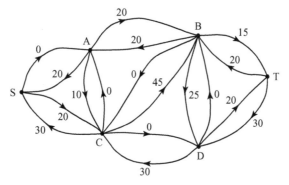

Figure 5.16

So, for example, because you are now increasing the flow along AB by 20, the scope for a further increase along this arc has reduced from 40 to 20, whilst the scope for decreasing the flow along this arc has risen from 0 to 20.

In general:

■ when the flow along an arc is increased, the amount is subtracted from the value on the forward arrow for that arc and added to the value on the backward arrow

■ the forward arrow shows the spare capacity of an arc; i.e. the amount by which the flow along it can be increased

■ the backward arrow shows the current flow along an arc, which is the amount by which the flow can be decreased

■ both forward and backward arrows can be used to find a route, but there must be a positive amount in the direction in which you want to change the flow.

At this point the algorithm diverges from the earlier method by inspection, as it doesn't allow for one path being replaced by another. However, you can easily spot an additional flow of 20 along the path SCBDT. The new network of potential changes is shown in Figure 5.17.

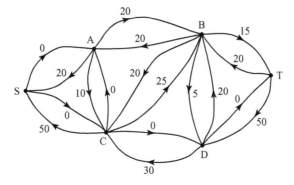

Figure 5.17

The arcs leading from S are both of capacity zero, and this means that no further improvements are possible.

The flows that are introduced during the process are often displayed in a table, as in Table 5.1.

Route	Flow
SCDT	30
SABT	20
SCBDT	20
Total	70

Table 5.1

This agrees with the total flow of 70, found previously. However, the combined effect of these flows, as shown in Figure 5.18, is different from that found by inspection.

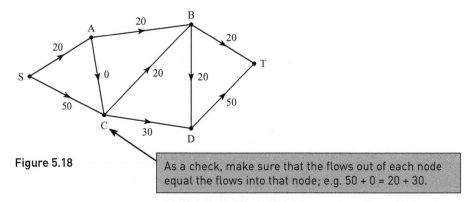

Figure 5.18

As a check, make sure that the flows out of each node equal the flows into that node; e.g. 50 + 0 = 20 + 30.

The flow augmenting procedure can sometimes involve **backflows**. This occurs when the flow that was initially chosen along a particular arc needs to be reduced (or removed entirely), in order to re-route the flow along a different arc and thereby increase the overall flow. This is illustrated in the next example.

Example 5.2

Figure 5.19 shows the capacity for each arc of a network and Figure 5.20 shows an initial flow for this network.

Show that the total flow can be increased to 14.

Solution

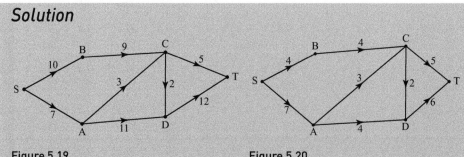

Figure 5.19   **Figure 5.20**

Figure 5.20 shows an initial flow that has been chosen by inspection, so that the flow along each arc doesn't exceed its capacity, and the total of the flows leading from each node is equal to the total of the flows leading into it. The total flow across the network is 11.

Figure 5.21 shows the network of potential changes. For example, for the arc leading from S to B, the 'forward' flow of 6 brings you up to the maximum capacity of 10, whilst the 'backward' flow of 4 is the most that the flow could be reduced by (being the current flow along the arc).

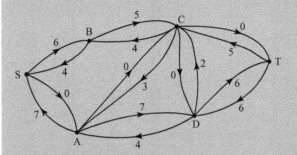

Figure 5.21

Note that there is no scope for increasing the flow from S to A. But you can follow the route SBCADT: an additional flow of 3 is possible. This involves a backflow along CA. After making the change, you obtain Figure 5.22.

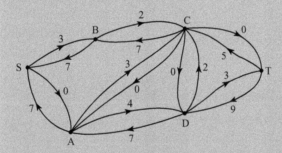

Figure 5.22

All the arcs leading from C now have zero value, and so no further improvements can be made. Figure 5.23 shows the new version of the actual flow. The total flow across the network has increased to 14.

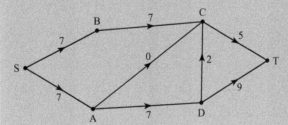

Figure 5.23

Exercise 5.2

① A gas distribution network consisting of three supply points, A, B and C, three intermediate pumping stations, P, Q and R, and two delivery points, X and Y, as well as connecting pipes, is shown in Figure 5.24.

The figures by A, B and C are measures of the daily availability of gas at the supply points.

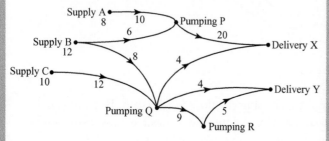

Figure 5.24

(i) Introduce a single source and a single sink and draw the revised network. The links from the source should reflect the supply availabilities.

(ii) By inspection, find the maximum daily flow through the network, listing the flows in each pipe.

(iii) Prove that your flow is a maximum.

② For the network in question 1 of Exercise 5.1

(i) use flow augmentation to find a flow pattern that gives a maximum total flow

(ii) prove that your total flow is a maximum.

③ For the network in question 3 of Exercise 5.1, use flow augmentation to find a flow pattern that gives a maximum total flow.

④ Gas is supplied to three locations, T_1, T_2 and T_3, from two sources S_1 and S_2. The rate of supply from S_1 cannot exceed 20 units. The rate of supply from S_2 cannot exceed 30 units. There are no constraints on the rate of flow into T_1, T_2 and T_3. The transmission network is shown in Figure 5.25, pipe capacities giving the maximum permissible rates of flow.

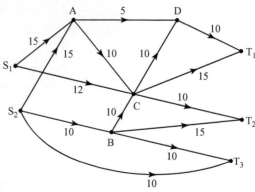

Figure 5.25

(i) Add a supersource together with appropriate capacitated arcs to model the supply constraints.

(ii) Add a supersink together with appropriate arcs.

(iii) Find the maximum total flow through the network, saying how much is delivered to each of T_1, T_2 and T_3.

⑤ The diagram below shows the capacities of the edges in a directed distribution network and the flows currently established in the network. The capacity of each pipe is given by the number not circled on each arc. The number inside the circle on each arc represents the actual flow.

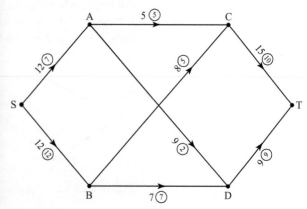

Figure 5.26

(i) Find a minimum cut and give its capacity.

(ii) Say why the established flows do not give a maximum total flow through the network.

(iii) Find and give a flow-augmenting path, and give a set of flows that do produce a maximum flow through the network.

⑥ Figure 5.27 shows a directed flow network with capacities and established flows. The capacity of each pipe is given by the number not circled on each arc. The number inside the circle on each arc represents the actual flow.

(i) Use flow augmentation to find the maximum flow.

(ii) Explain how the cut marked on the diagram confirms that the flow you have found is a maximum.

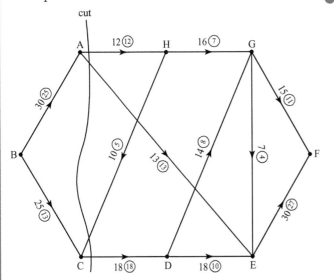

Figure 5.27

5 Refinements

Nodes of restricted capacity

It is possible that an upper limit exists for the flow that can pass through a particular node. This can be dealt with by replacing the node with two (unrestricted) nodes, connected by an arc of capacity equal to the limit placed on the original node. Figure 5.28 shows the network in Figure 5.1, where there is now a limit of 20 placed on the flow through node B.

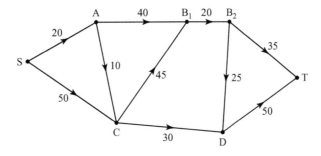

Figure 5.28

Arcs with lower capacities

In some situations, the flow for a particular arc may not be allowed to fall below a certain value. When labelling the original network, you now need to show lower and upper capacities, as for example in Figure 5.29 (which also shows some possible cuts).

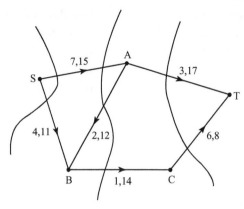

Figure 5.29

As before, you can try to maximise the flow through the network by inspection. Finding a cut of the same value will confirm the maximum flow.

ACTIVITY 5.1

Maximise the flow across the network shown in Figure 5.29 by inspection.

When determining the value of a cut, you can no longer exclude any arcs where the flow is towards the source. Instead, the lower capacities of these arcs are subtracted from the total maximum flow leading away from the source. The example of the border between two countries should make this clear: if it is a requirement that there is some traffic going back towards the source along a particular arc, then the amount of this traffic will reduce the net flow across the border in the direction of the sink.

ACTIVITY 5.2

Find a cut with a value of 21 in Figure 5.29.

Applying this to Figure 5.29, the cut {S}/{A,B,C,T} has value $15 + 11 = 26$; the cut {S,B}/{A,C,T} has value $15 - 2 + 14 = 27$, and the cut {S,A,B,C}/{T} has value $17 + 8 = 25$.

This means that the maximum flow cannot exceed 25.

Example 5.3

Apply the flow augmenting process to the network shown in Figure 5.29.

Solution

The procedure is the same as used previously, except that the initial flow chosen must take account of any lower capacities – as must the potential increases and decreases for the arcs.

Suppose that you start with the following flow.

SABCT 4

SAT   11

SBCT   4,

with a total of 19.

This gives rise to the network of potential changes in Figure 5.30.

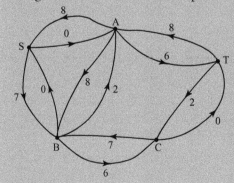

ACTIVITY 5.3

Identify a flow through the network of potential changes in Figure 5.30.

Figure 5.30

Exercise 5.3

① Find a flow pattern that gives a maximum total flow for the network in Figure 5.28.

② Continuing question 1 of Exercise 5.2, a new pipeline is proposed, with a capacity of 5 units per day, connecting P and Q.

   (i) Apply the flow augmenting procedure to find the new maximum flow.

   (ii) R is now a delivery point. Explain how to adapt the approach used in Ex 5.2 Q1 (i) to find the maximum daily flow of gas in total that can be delivered to R, X and Y. You do not need to calculate this flow.

③ Find a flow pattern that gives a maximum total flow for the network in Figure 5.31.

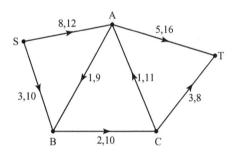

Figure 5.31

④ The matrix below represents capacities of roads (thousands of cars per hour) in a road network.

	A	B	C	D	E	F	G
A	–	3	3	–	–	2	–
B	3	–	–	5	1	–	–
C	3	–	–	2	–	3	–
D	–	5	2	–	–	1	2
E	–	1	–	–	–	–	1
F	2	–	3	1	–	–	2
G	–	–	–	2	1	2	–

Figure 5.32

   (i) Draw the network.

   (ii) Find the maximum hourly flow of cars from B to F, showing how this may be achieved. Prove that this is a maximum.

⑤ The diagram below shows a pipe network. The numbers on the arcs give the capacities of the pipes.

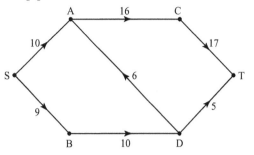

Figure 5.33

   (i) Find the maximum flow through the network, showing the steps in your working.

   (ii) Prove that your flow is a maximum.

⑥ The network below shows a system of pipes. The numbers represent the capacities of the pipes.

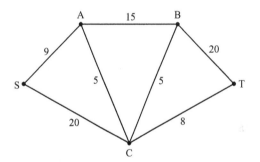

Figure 5.34

   (i) Show that a flow of 27 units from S to T is possible.

   (ii) Find a cut with capacity 27 units. Explain its significance.

   (iii) If there must be a flow of at least 2 from A to C in AC what will be the maximum flow from S to T, and what flow will be needed in SC to achieve that maximum?

⑦ In the following network the arcs have capacities in the directions indicated by the arrows. The capacity of each pipe is given by the number not circled on each arc. The number inside the circle on each arc represents the actual flow.

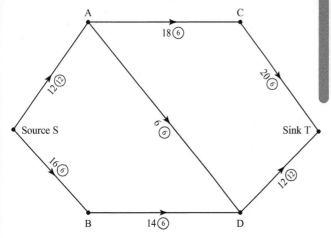

Figure 5.35

(i) Find the minimum cut and explain why it shows that the existing flow is not a maximum flow. Show the flows along the arcs that give the maximum flow from S to T.

Suppose that the flow along the arc AD may now be in either direction.

(ii) Explain why the maximum flow from S to T is now greater than in part (i) and state the value of the new maximum. Show the flows along the arcs that give that maximum flow from S to T.

KEY POINTS

1 The aim is to find the maximum possible flow through a network, from a source to a sink.
2 A cut divides a network into two parts, separating the source and the sink.
3 The value of a cut places a limit on the maximum flow possible, and this gives rise to the maximum flow–minimum cut theorem.
4 Supersources and supersinks may be introduced to deal with multiple sources and sinks.
5 The flow augmenting algorithm may be employed to identify the maximum possible flow.
6 Adjustments can be made to allow for nodes of restricted capacity and arcs with lower capacities.

6

Game theory

→ **Matching pennies** is the name for a simple game used in **game theory**. It is played between two players, Even and Odd. Each player has a penny and must secretly turn the penny to heads or tails. The players then reveal their choices simultaneously. If the pennies match (both heads or both tails), then Even keeps both pennies, so wins one from Odd (+1 for Even, –1 for Odd). If the pennies do not match (one heads and one tails), Odd keeps both pennies, so receives one from Even (–1 for Even, +1 for Odd).

1 Pay-off matrices

Game theory aims to model situations involving competitors (or **players**), where each player's gain (or loss) depends not only on their decisions, but also the decisions of their competitors. It can be applied in areas such as business negotiations, election campaigns, and warfare.

This chapter is about **zero-sum** games between two players. This means that if player 1 wins a certain amount, then player 2 loses that amount. For example, suppose that both players have two options: A and B. The amount (in £) won by player 1 (and lost by player 2) is given by the **pay-off matrix** in Table 6.1.

	Player 2 plays A	Player 2 plays B
Player 1 plays A	1	0
Player 1 plays B	−5	10

The pay-off matrix shows player 1's gains and player 2's losses.

Table 6.1

The elements in the pay-off matrix are given from player 1's point of view (this is the standard convention).

If player 1 wants to have a chance of winning £10, they must choose option B, and accept the risk of losing £5 if player 2 chooses option A, The safer course of action is to choose option A and either win £1 or break even.

Because this is a zero-sum game, to consider things from player 2's point of view, you transpose the matrix and reverse the signs of the elements in the matrix, to give the matrix in Table 6.2.

	Player 1 plays A	Player 1 plays B
Player 2 plays A	−1	5
Player 2 plays B	0	−10

Table 6.2 (player 2's gains)

The pay-off matrix now shows player 2's gains and player 1's losses. The values are the same but the signs are reversed because it is a zero-sum game.

The matrix is transposed (i.e. rows become columns and vice versa) so that the rows now give player 2's gains and the columns give player 1's losses.

Option A offers player 2 the chance of winning £5, with the risk of losing £1, and choosing option B means either breaking even or losing £10.

The following factors may affect the choices made by each player.

- How risk averse is the player? Player 1 may be able to afford to lose £5, but the situation might be different for £5000.

- Will the game be repeated many times? A player may then be able to bear an occasional loss.

- Can the choice of the other player be predicted? For example, if the game is repeated, the other player may always make the same choice.

Predicting the other player's choice is of particular interest in game theory.

For the current example, player 1 may conclude that player 2 is more likely to choose option A, and so player 1 may decide to choose option A themselves, and win £1 (rather than lose £5). Player 2 will then lose £1.

If player 2 discovers player 1's intention to choose option A, then it is better for player 2 to choose option B, and so break even.

2 Play-safe strategies

A **play-safe strategy** is where a player chooses the option with the best worst outcome. In other words, they assume that the other player penalises them the most by their choice.

In the previous example, if both players adopt a play-safe strategy, player 1 chooses option A (which has a worst outcome of £0, compared to the worst outcome of −£5 for option B), and player 2 also chooses option A (which has a worst outcome of −£1, compared to the worst outcome of −£10 for option B).

The result is that player 1 wins £1, and player 2 loses £1. It is assumed that both players have to play the game.

Now consider the pay-off matrix in Table 6.3, where player 1 has two choices and player 2 has three.

	Player 2 plays A	Player 2 plays B	Player 2 plays C
Player 1 plays A	2	−1	1
Player 1 plays B	1	−2	−3

Table 6.3

As usual, these are the pay-offs for player 1. The pay-offs for player 2 are shown in Table 6.4.

	Player 1 plays A	Player 1 plays B
Player 2 plays A	−2	−1
Player 2 plays B	1	2
Player 2 plays C	−1	3

Table 6.4 (player 2's gains)

The matrix has been transposed and the signs have been reversed.

The worst outcomes for player 1 can be shown in an additional column, as in Table 6.5 with the better option (from player 1's point of view) circled.

	Player 2 plays A	Player 2 plays B	Player 2 plays C	Worst
Player 1 plays A	2	–1	1	(–1)
Player 1 plays B	1	–2	–3	–3

Table 6.5

The worst outcomes for player 2 can also be shown in an additional column, as in Table 6.6, where the best option (from player 2's point of view) is circled.

	Player 1 plays A	Player 1 plays B	Worst
Player 2 plays A	–2	–1	–2
Player 2 plays B	1	2	(1)
Player 2 plays C	–1	3	–1

Table 6.6 (player 2's gains)

The play-safe strategy for player 1 is option A, whilst for player 2 it is option B.

If both players adopt the play-safe strategy, player 1 will not want to change their choice (assuming that player 2 doesn't change from option B), as –1 is better than –2.

Similarly, player 2 will not want to change their choice (assuming that player 1 doesn't change from option A), as 1 is better than both –2 and –1.

This is known as a **stable solution**. It is also called a **saddle point**, and the game is said to be in equilibrium. The stable solution exists because, in player 1's matrix, –1 is below both 2 and 1 (in row A), whilst –1 is above –2 (in column B).

So, where a stable solution exists, it is in both players' interests for them to adopt the play-safe strategy. The **value of the game** is defined as player 1's pay-off.

Although you considered player 2's matrix in the previous discussion, the standard procedure for identifying a stable solution is based on player 1's matrix.

You establish the worst outcomes for player 1, and place these in a column to the right of the pay-off matrix. These are called the **row minima** (i.e. the column shows the minimum for each row).

Because the elements of the matrix are now player 2's outcomes with the sign reversed, the column of worst outcomes for player 2 now becomes the row of maximum values for each of the columns. These are called the **column maxima**.

This new system of labelling is shown in Table 6.7.

	A	B	C	Row minima
A	2	–1	1	(–1)
B	1	–2	–3	–3
Column maxima	2	(–1)	1	

Table 6.7

This is called the min (column maxima).

This is called the max (row minima).

A stable solution occurs because –1 is the minimum value in row A, and also the maximum value in column B.

In general, a stable solution occurs when the maximum of the row minima equals the minimum of the column maxima.

> **Note**
>
> In 3D geometry, a point on a surface is a saddle point if it is a minimum point of the surface in one direction, but a maximum in another.

> **Note**
>
> So far, you have used an additional matrix from player 2's point of view. It is conventional to work only with the matrix from player 1's point of view, and this will be done from now on.
>
> Unless stated otherwise, this means that the matrices will typically just have A, B, etc. in the column headings rather than the name of the strategy, 'Player 2 plays A', etc. The same goes for the 'Player 1 plays ...' entries in the row headings.

A zero-sum game has a stable solution if and only if

the max (row minima) = the min (column maxima)

If a stable solution exists, it is the best strategy for both players.

Exercise 6.1

① The pay-off matrix for a zero-sum game between players 1 and 2 is as follows.

	A	B	C
A	2	0	1
B	−1	3	−4

Table 6.8

(i) Determine the pay-off matrix from player 2's point of view.
(ii) What is the most that player 2 can win?
(iii) What is the most that player 2 can lose?

② The pay-off matrix for a zero-sum game between players 1 and 2 is as follows.

	A	B	C
A	0	1	1
B	−1	5	2
C	4	−2	−3

Table 6.9

(i) What will be the outcome if both players play safe?
(ii) What will be the outcome if player 1 plays safe, and player 2 hears of player 1's intention?
(iii) What will be the outcome if (instead) player 2 plays safe, and player 1 hears of player 2's intention?

③ The pay-off matrix for a zero-sum game between players 1 and 2 is as follows.

	A	B	C
A	4	−1	2
B	0	−3	−2
C	−3	−2	6
D	−5	−4	3

Table 6.10

Show that this game has a stable solution, and give the value of the game.

④ For each of the following pay-off matrices for zero-sum games, determine the play-safe strategies for player 1 and player 2.

(i)

	A	B
A	1	−1
B	2	0

Table 6.11

(ii)

	A	B	C	D
A	3	2	0	1
B	1	−2	−1	0

Table 6.12

(iii)

	A	B	C
A	−2	1	4
B	0	2	−1
C	3	−3	0

Table 6.13

(iv)

	A	B
A	1	3
B	2	0
C	4	−1

Table 6.14

⑤ For the pay-off matrices in question 4, identify any stable solutions and give the values of the games in these cases.

⑥ Show that if $a < b < c < d$, then a stable solution will exist for the zero-sum game pay-off matrix shown in Table 6.15.

	A	B
A	a	b
B	d	c

Table 6.15

3 Dominated strategies

Where a stable solution doesn't exist, it may be possible to devise a strategy for both players where neither player's choices are predictable. This is referred to as a **mixed strategy**, and is considered in the next section.

However, before adopting this strategy, there is one simplifying process that can often be applied. This involves the idea of **dominance**.

	A	B	C
A	1	2	2
B	−2	1	2
C	3	−1	0

Table 6.16

Referring to the pay-off matrix in Table 6.16 (which, by convention, is from player 1's point of view), you see that player 2 would never choose option C, because option B would always be either better or of equal value (from player 2's point of view), whatever choice player 1 made. Column B is said to **dominate** column C, and the pay-off matrix can be reduced to that in Table 6.17.

	A	B
A	1	2
B	−2	1
C	3	−1

Table 6.17

Similarly, for the new matrix in Table 6.17, player 1 would never choose option B, as row A dominates row B, and you can reduce the matrix further to that in Table 6.18.

	A	B
A	1	2
C	3	−1

Table 6.18

As will be seen in Exercise 6.2, using the idea of dominance sometimes leads to a 1×1 matrix, which gives the value of the game directly.

Exercise 6.2

① Consider the following pay-off matrix.

	A	B	C	D
A	2	1	3	3
B	0	0	−1	0
C	1	1	0	0

Table 6.19

(i) Which option(s) should player 1 never choose?
(ii) Which option(s) should player 2 never choose?

② For each of the pay-off matrices in question 4 of Exercise 6.1, identify any dominated strategies, and give the reduced matrix in these cases.

③ (i) Show that there is a stable solution for the following pay-off matrix, without reducing the matrix. Give the value of the game.

	A	B	C	D	E
A	1	0	−1	2	3
B	1	2	0	1	2
C	2	1	−1	4	3

Table 6.20

(ii) Reduce the pay-off matrix, as far as possible.

④ Reduce the following pay-off matrix as much as possible, and show that there is no stable solution to the game.

	A	B	C	D
A	2	−1	0	4
B	0	6	2	2
C	4	0	1	1
D	1	−3	−1	2
E	−2	3	1	−3

Table 6.21

⑤ For what values of x can the following pay-off matrix be reduced?

	A	B	C
A	2	x	1
B	1	4	0

Table 6.22

4 Optimal mixed strategies

In a mixed strategy the players select their options with a certain probability, so that their expected outcomes are optimised. Because the games are zero-sum, the expected outcome of one player is the expected outcome of the other, with the sign reversed.

A mixed strategy can be employed whenever the pay-off matrix is of order $2 \times n$ (where $n \geqslant 2$), after the matrix has been reduced due to any dominance.

You start by looking at the 2×2 case.

Example 6.1

Establish whether there is a stable solution for the zero-sum game given by the following pay-off matrix.

	A	B	Row minima
A	1	2	①
C	3	−1	−1
Column maxima	3	②	

Table 6.23

Solution

Max (row minima) = 1

Min (column maxima) = 2

Compare the max (row minima) with the min (column maxima).

As these values are not equal, there is not a stable solution.

> ### ■ Note
> --
> In Table 6.23, you see that if player 1 plays safe, they choose option A, as this guarantees that the worst outcome is 1 (as opposed to a worst outcome of −1 if option C is chosen).
>
> Similarly, player 2 will choose option B, as this guarantees a worst outcome of −2 (2 for player 1), compared with the worst outcome of −3 if option A is chosen.
>
> This means that the actual outcome is 2 for player 1 and −2 for player 2.
>
> If player 1 knows that player 2 will always choose option B, then they are happy to choose option A (as 2 ⩾ −1). But if player 2 knows that player 1 will always choose option A, then player 2 will want to change their choice to option A (as 1 < 2, or −1 > −2 from player 2's point of view).
>
> But if player 1 knows that player 2 will choose option A, then player 1 will now want to choose option C, and so on. It can be shown that a stable solution is never obtained.

The following problems occur when there is no stable solution.

■ Neither player can safely assume what the other player's choice will be.

■ Where the game is repeated, one player may take advantage of the other, if their choice becomes predictable.

It is possible to find a strategy for each player that avoids these problems, by deliberately introducing an element of uncertainty, so that each player makes their choice with certain probabilities. This is the mixed strategy already mentioned.

For the 2×2 matrix in Table 6.23, suppose that player 1 adopts option A with probability p, and option C with probability $1 - p$.

Player 1's expected pay-off depends on player 2's choice.

If player 2 chooses option A, then player 1's expected pay-off is

$$1p + 3(1 - p) = 3 - 2p.$$

If, instead, player 2 chooses option B, then player 1's expected pay-off is

$$2p + (-1)(1 - p) = 3p - 1.$$

These expected pay-offs are functions of p, and can be shown graphically, as in Figure 6.1.

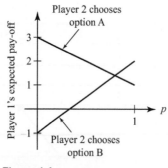

Figure 6.1

You see that player 1 can maximise their expected pay-off (obtaining 3) when player 2 chooses option A, by setting $p = 0$. However, they would risk receiving a pay-off of only −1 if player 2 chose option B. Similarly, if player 2 chooses option B, player 1's expected gain is maximised (obtaining 2) by setting $p = 1$, but they risk obtaining only 1 if player 2 chooses option A instead.

The safest strategy is to assume that player 2's choice always results in the worst possible outcome for player 1, so that whatever you set p equal to, player 1's expected pay-off is given by the lower of the two lines.

This means that you need to look for the value of p such that the lower of the two lines is as high as possible, and this occurs where the two lines intersect.

Example 6.2

Find the required value of p for the zero-sum game shown by the pay-off matrix in Table 6.23.

Note

It does not matter whether the game is repeated: the expected pay-off for each game is 1.4, regardless of the number of games played.

Solution

$$3 - 2p = 3p - 1 \Rightarrow 4 = 5p \Rightarrow p = 0.8$$

> To find where the two lines intersect you need to solve their equations simultaneously.

With this value for p, player 1 does not mind what player 2 does: if the game is repeated, player 2 might choose the same option all the time, or they might themselves choose their option according to a probability rule.

Player 1's expected pay-off is $3 - 2p$ (or $3p - 1$) $= 1.4$

Example 6.3

Find the probability rule that player 2 should adopt, and their expected pay-off.

Solution

Suppose that player 2 adopts option A with probability q, and option B with probability $1 - q$.

If player 1 chooses option A, then player 2's expected pay-off is

$$-\{1q + 2(1 - q)\} = q - 2$$

> The minus sign is needed because the values in the pay-off matrix show player 2's losses, not gains.

If, instead, player 1 chooses option C, then player 2's expected pay-off is

$$-\{3q + (-1)(1 - q)\} = 1 - 4q$$

It is not essential to draw the graphs. However, they are shown in Figure 6.2.

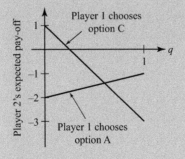

Figure 6.2

Note

Where a mixed strategy is employed, the value of the game is defined to be player 1's expected pay-off; in this case, 1.4.

You then find the required value of q from $q - 2 = 1 - 4q \Rightarrow 5q = 3 \Rightarrow q = 0.6$

Player 2's expected pay-off is $q - 2$ (or $1 - 4q$) $= -1.4$

As this is a zero-sum game, it is no surprise that player 2's expected pay-off is just player 1's expected pay-off with the sign reversed.

$2 \times n$ and $n \times 2$ matrices

This method can be extended to $2 \times n$ matrices; i.e. where player 2 has n choices.

In the case of $n \times 2$ matrices, where player 1 has n choices, the method cannot be used to find player 1's probability rule, as this would require more than one probability variable (instead of the single p).

However, the value of the game can be established by swapping the roles of player 1 and player 2, to obtain a $2 \times n$ matrix, and then find player 1's probability rule.

Example 6.4

Find expressions, in terms of p, for player 1's expected pay-off, for each of player 2's three options, for the pay-off matrix in Table 6.24.

Draw the graphs of the corresponding straight lines on a single diagram.

	A	B	C
A	0	−1	2
B	2	3	−2

Table 6.24

Solution

The first step is always to check if the matrix can be reduced, due to any dominance of either rows or columns. Here there is no dominance.

As before, suppose that player 1 adopts option A with probability p and option B with probability $1 - p$.

If player 2 chooses option A, then player 1's expected pay-off is

$0p + 2(1 - p) = 2 - 2p$

If player 2 chooses option B, then the expected pay-off is

$(-1)p + 3(1 - p) = 3 - 4p$

If player 2 chooses option C, the expected pay-off is

$2p + (-2)(1 - p) = 4p - 2$

The graphs of the three straight lines are shown in Figure 6.3.

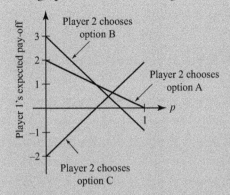

Figure 6.3

Once again, you need to look for the value of p such that the lower of the lines is as high as possible. This occurs where the lines for options B and C intersect. Note that it was necessary to draw the graphs in order to see this.

Hence $3 - 4p = 4p - 2$, so that $5 = 8p$ and $p = \frac{5}{8}$.

With this value of p, the expected pay-off for player 1 is $3 - 4p$ (or $4p - 2$) $= \frac{1}{2}$ (this is the value of the game).

The following example shows a device that can be used to find player 2's probability rule, where player 2 has three options.

Example 6.5

Find player 2's probability rule for the pay-off matrix (Table 6.24) of Example 6.4.

Solution

Suppose that player 2 adopts options A, B and C with probabilities q, r and $1 - q - r$, respectively.

If player 1 chooses option A, then player 2's expected pay-off is

> The minus sign is needed because the values in the pay-off matrix show player 2's losses, not gains.

$$-\{0q + (-1)\,r + 2\,(1 - q - r)\} = -2 + 2q + 3r$$

If, instead, player 1 chooses option B, then player 2's expected pay-off is

$$-\{2q + 3r + (-2)\,(1 - q - r)\} = 2 - 4q - 5r$$

As usual, the probability rule is chosen in such a way that the expected pay-offs are the same, whichever option player 1 chooses, and both are equal to the value of the game from player 2's point of view. (This corresponds to the intersection of the lines in the simple 2×2 case.)

So $-2 + 2q + 3r = 2 - 4q - 5r = -\frac{1}{2}$ (as the value of the game from player 1's point of view is $\frac{1}{2}$, from Example 6.4).

These equations can be written as

> Remember to always state the probabilities at the end of your calculation.

$-4 + 4q + 6r = -1$ and $4 - 8q - 10r = -1$ or $4q + 6r = 3$ ① and $8q + 10r = 5$ ②.

$2 \times$ ① $-$ ② gives $2r = 1$, so that $r = 0.5$, and $q = 0$

Thus the probability rule for player 2 is to never choose option A and choose options B and C with equal probability (0.5).

① Show that there is no stable solution for the following pay-off matrix.

	A	B
A	3	2
B	1	4

Table 6.25

Use a mixed strategy to find the value of the game.

② (i) Consider the game with the following pay-off matrix.

	A	B
A	1	2
B	3	0
C	−1	0
D	−1	4

Table 6.26

Are there any choices that either player should never make?

(ii) Find the value of the game.

③ Use a mixed strategy to find the values for player 1 of the games having the following pay-off matrices.

(i)

	A	B
A	1	4
B	5	3

Table 6.27

(ii)

	A	B	C
A	−1	2	5
B	9	7	3

Table 6.28

(iii)

	A	B
A	−1	5
B	2	4
C	7	3

Table 6.29

④ Determine optimal strategies for both players for the game with the following pay-off matrix, and draw a graph to illustrate the situation from player 1's point of view.

	A	B
A	3	−1
B	2	4

Table 6.30

⑤ Determine the value of the game with the following pay-off matrix.

	A	B	C
A	5	2	−3
B	−1	4	6

Table 6.31

⑥ Determine optimal strategies for both players for the game with the following pay-off matrix, and establish the value of the game.

	A	B	C
A	4	3	2
B	−2	1	4
C	2	1	5

Table 6.32

5 Converting games to linear programming problems

So far, you have established methods for dealing with the following situations.

- 2×2 pay-off matrices
- cases where there is dominance, so that the pay-off matrix can be reduced
- $2 \times n$ or $n \times 2$ pay-off matrices.

For other orders of matrices, you can turn to linear programming methods.

In order to illustrate the method, consider the pay-off matrix in Table 6.33.
This uses the same numbers as in Table 6.24 (swapping the roles of the two players in Table 6.24), and the value of the game can be found to be $\frac{1}{2}$.

Note

This is a new example, where player 1 now has three options.

	A	B
A	0	2
B	−1	3
C	2	−2

Table 6.33

Suppose that player 1 chooses the three options with probabilities p_1, p_2 and p_3.

Let v be the value of the game.

In order for the Simplex method to work, you require the control variables to be non-negative. The probabilities are non-negative, but you also want $v \geqslant 0$. Suppose that the pay-off matrix is of the general form shown in Table 6.34, and that any dominance has been removed.

	A	B
A	λ_1	μ_1
B	λ_2	μ_2
C	λ_3	μ_3

Table 6.34

Then, as before, v will be found from the intersection of $\lambda_1 p_1 + \lambda_2 p_2 + \lambda_3 p_3$ and $\mu_1 p_1 + \mu_2 p_2 + \mu_3 p_3$,

so that $v = \lambda_1 p_1 + \lambda_2 p_2 + \lambda_3 p_3 = \mu_1 p_1 + \mu_2 p_2 + \mu_3 p_3$ (*)

In order for the Simplex method to work, you need to have positive values for the λ_i and μ_i. The reason for this will be explained shortly.

As it is only the relative values of the elements that are important, you can add any amount to all of the elements, provided you subtract it from the expected pay-off. For the present example, you can add 3 to each element, to produce the matrix in Table 6.35.

	A	B
A	3	5
B	2	6
C	5	1

Table 6.35

Using this new matrix, the expected pay-offs for player 1 are $3p_1 + 2p_2 + 5p_3$ or $5p_1 + 6p_2 + 1p_3$, depending on whether player 2 chooses option A or option B.

As before, assume the worst case scenario; namely that, whatever probabilities are adopted, player 2's choice will always result in v being the lower of $3p_1 + 2p_2 + 5p_3$ and $5p_1 + 6p_2 + 1p_3$. Given this constraint, aim to maximise v.

Also, $p_1 + p_2 + p_3 = 1$, but for the purpose of linear programming an inequality is required. However, it is possible to write $p_1 + p_2 + p_3 \leq 1$, for the following reason:

Suppose that v has been maximised with $p_1 + p_2 + p_3 < 1$. Then it would be possible to increase one of the probabilities, and thereby increase both $3p_1 + 2p_2 + 5p_3$ and $5p_1 + 6p_2 + 1p_3$, so that the lower of these expressions would be increased, giving a higher value for v, which contradicts the assumption that v has been maximised. So $p_1 + p_2 + p_3 < 1$ isn't possible, and therefore $p_1 + p_2 + p_3 = 1$.

Note that, in the above argument, when one of the probabilities is increased, the expressions for v (in general, $\lambda_1 p_1 + \lambda_2 p_2 + \lambda_3 p_3$ and $\mu_1 p_1 + \mu_2 p_2 + \mu_3 p_3$) can only be guaranteed to increase if the λ_i and p_i are all positive.

The linear programming problem can therefore be expressed as

maximise $p = v - 3$ subject to

$$v \leq 3p_1 + 2p_2 + 5p_3$$
$$v \leq 5p_1 + 6p_2 + 1p_3$$
$$p_1 + p_2 + p_3 \leq 1$$
$$p_1, p_2, p_3 \geq 0.$$

This can be solved using the Simplex method.

Example 6.6

Use the Simplex method to find the value of the game with the following pay-off matrix.

	A	**B**
A	1	2
B	3	−1

Table 6.36

> You have already found this value to be 1.4 in Example 6.2.

Solution

Add 2 to each element, so that all elements are positive, to give the following matrix.

	A	**B**
A	3	4
B	5	1

Table 6.37

Maximise $P = v - 2$, subject to the constraints

$$v \leq 3p_1 + 5p_2, v \leq 4p_1 + p_2, p_1 + p_2 \leq 1 \ (p_1, p_2 \geq 0, v > 0).$$

Introducing slack variables, the Simplex equations are

$$P - v = -2$$
$$v - 3p_1 - 5p_2 + s_1 = 0$$
$$v - 4p_1 - p_2 + s_2 = 0$$
$$p_1 + p_2 + s_3 = 1$$

The Simplex tableau is shown in Table 6.38.

P	v	p_1	p_2	s_1	s_2	s_3	Value	Equation
1	-1	0	0	0	0	0	-2	①
0	①	-3	-5	1	0	0	0	②
0	1	-4	-1	0	1	0	0	③
0	0	1	1	0	0	1	1	④
1	0	-3	-5	1	0	0	-2	⑤ = ① + ⑥
0	1	-3	-5	1	0	0	0	⑥ = ②
0	0	-1	④	-1	1	0	0	⑦ = ③ − ⑥
0	0	1	1	0	0	1	1	⑧ = ④
1	0	$-\dfrac{17}{4}$	0	$-\dfrac{1}{4}$	$\dfrac{5}{4}$	0	-2	⑨ = ⑤ + 5 × ⑪
0	1	$-\dfrac{17}{4}$	0	$-\dfrac{1}{4}$	$\dfrac{5}{4}$	0	0	⑩ = ⑥ + 5 × ⑪
0	0	$-\dfrac{1}{4}$	1	$-\dfrac{1}{4}$	$\dfrac{1}{4}$	0	0	⑪ = ⑦ ÷ 4
0	0	$\left(\dfrac{5}{4}\right)$	0	$\dfrac{1}{4}$	$-\dfrac{1}{4}$	1	1	⑫ = ⑧ − ⑪
1	0	0	0	$\dfrac{3}{5}$	$\dfrac{2}{5}$	0	$\dfrac{7}{5}$	⑬ = ⑨ + $\dfrac{17}{4}$ × ⑯
0	1	0	0	$\dfrac{3}{5}$	$\dfrac{2}{5}$	0	$\dfrac{17}{5}$	⑭ = ⑩ + $\dfrac{17}{4}$ × ⑯
0	0	0	1	$-\dfrac{1}{5}$	$\dfrac{1}{5}$	$\dfrac{1}{5}$	$\dfrac{1}{5}$	⑮ = ⑪ + $\dfrac{1}{4}$ × ⑯
0	0	1	0	$\dfrac{1}{5}$	$-\dfrac{1}{5}$	$\dfrac{4}{5}$	$\dfrac{4}{5}$	⑯ = ⑫ × $\dfrac{4}{5}$

Table 6.38

The maximised value of P is $\dfrac{7}{5} = 1.4$, as before.

Also, as a check, $p_1 + p_2 = 1$, where $p_1 = \dfrac{4}{5}$ and $p_2 = \dfrac{1}{5}$, also as before.

Exercise 6.4

① Formulate the linear programming problem that can be used to find the value of the game with the following pay-off matrix.

	A	B
A	1	3
B	2	0
C	4	−1

Table 6.39

② Use the Simplex method to find the value of the game in question 1 of Exercise 6.3, with the following pay-off matrix.

	A	B
A	3	2
B	1	4

Table 6.40

③ Use the Simplex method to find the values of the games in question 3 of Exercise 6.3.

④ Formulate the linear programming problem that can be used to find the value of the game with the following pay-off matrix. Write down the initial Simplex equations.

	A	B
A	−2	−1
B	−3	0
C	1	−2

Table 6.41

⑤ Establish the initial Simplex tableau that can be used to find the value of the game with the following pay-off matrix.

	A	B
A	4	−3
B	−1	2

Table 6.42

⑥ For what value(s) of x is there a stable solution for the following pay-off matrix?

	A	B
A	2	x
B	0	3
C	1	2

Table 6.43

⑦ A mathematical board game for two players is being trialled. When player 1 lands on a particular square, the amount he or she wins is the highest common factor of two numbers. These numbers are chosen from 3 specified numbers, with each player choosing one of the numbers (they can both choose the same number). Player 2 is trying to minimise the amount that player 1 wins. Janet and John are playing this game, and Janet lands on a square with the numbers 24, 30 and 45.

(i) Which numbers should Janet and John choose if they adopt play safe strategies, and how much will Janet win in that case?

(ii) If (unknown to Janet) John has forgotten how to work out a highest common factor, and chooses his number at random, what amount can Janet expect to win, on average?

LEARNING OUTCOMES

Now you have finished this chapter, you should be able to

➤ understand, interpret and construct pay-off matrices

➤ find play-safe strategies and the value of the game

➤ prove the existence or non-existence of a stable solution

➤ identify and make use of dominated strategies

➤ find optimal mixed strategies for a game, including use of graphical methods

➤ convert higher-order games to linear programming problems, and solve them.

KEY POINTS

1 In a zero-sum game between two players, if player 1 wins a certain amount, then player 2 loses that amount.

2 The pay-off matrix shows the possible gains from player 1's point of view.

3 A play-safe strategy is where a player chooses the option with the smallest downside.

4 A stable solution exists when it is in both players' interests to adopt the play-safe strategy. This occurs when the maximum of the row minima equals the minimum of the column maxima.

5 A pay-off matrix can be reduced if one strategy dominates another.

6 In a mixed strategy, the players select their options with a certain probability in such a way that their expected outcomes are optimised.

7 The value of a game is player 1's pay-off, where there is a stable solution; or their expected pay-off when a mixed strategy is employed.

8 A mixed strategy can be employed whenever the (reduced) pay-off matrix is of order $2 \times n$ or $n \times 2$.

9 In more complicated cases, the Simplex method may be employed.

Binary operations and group theory

→ Place the numbers 1, 2 and 3 in the square below so that no number occurs more than once in any horizontal or vertical line.

1 Binary operations

A binary operation $*$ on a set S is a rule that assigns the element $x * y$ to any ordered pair of elements x, y in S.

Points to note

■ $x * y$ need not be an element of S, but if it is (in all cases), then the operation is said to be **closed**.

■ Symbols other than $*$ may be used ($\circ$ for example).

■ The set S can be finite (for example, the integers 0, 1, 2) or infinite (for example, $\mathbb{Z}^+$, the set of positive integers).

Note

The operations of addition, subtraction and multiplication are all binary operations on $\mathbb{Z}$ (the set of integers). However, division is not, as division by zero is undefined.

Modular arithmetic

The integers x and y are said to be **congruent modulo** m (or just **mod** m) if the difference between them is a multiple of m (and the symbol $\equiv$ is often used).

For example, $17 \equiv 52 \pmod 7$.

Every integer is congruent mod m to one of the numbers $0, 1, 2, ..., m - 1$.

Addition mod m ($+_m$) is defined on the set $\{0, 1, 2, ..., m - 1\}$ in the same way as ordinary addition, except that if the sum falls outside the set, then the answer is the remainder when the sum is divided by m. Multiplication mod m ($\times_m$) is defined in a similar way.

So $5 +_7 4 \equiv 2$ and $5 \times_7 4 \equiv 6$

The operations of $+_m$ and $\times_m$ on the set $\{0, 1, 2, ..., m - 1\}$ are both binary operations, as the results of the operations are defined for all pairs of elements in the set. They can also be shown to be closed.

Example 7.1

Determine whether the following operations on the given sets are binary operations. If so, are they closed?

- (i) $x * y = |x - y|$, where S is $\mathbb{R}$
- (ii) $+$ on $\mathbb{Z}^-$
- (iii) $x * y = \dfrac{x}{1 - y}$, where S is $\mathbb{R}$
- (iv)   Scalar product of vectors in two dimensions

Solution

- (i)   The operation is defined for all pairs of elements in the set, and so is a binary operation. The result is always a real number, so the operation is closed.

- (ii)   The sum of any two negative integers will be a negative integer, so the operation is a closed binary operation.

- (iii)   The operation is undefined for $y = 1$, and so is not a binary operation.

- (iv)   The scalar product is defined for all pairs of vectors in two dimensions, and so is a binary operation. The result is a scalar, and so the operation is not closed.

Further properties

A binary operation $*$ on S is said to be **commutative** if $a * b = b * a$ for all $a, b \in S$ (where the symbol $\in$ denotes 'is a member [or element] of').

A binary operation $*$ on S is said to be **associative** if

$(a * b) * c = a * (b * c)$, for all $a, b, c \in S$

Once an operation has been found to be associative, the brackets can be omitted.

Example 7.2

Are the following binary operations (a) commutative, and (b) associative?

- (i)   Matrix multiplication on the set of 2×2 matrices
- (ii) $x * y = |x - y|$, where S is $\mathbb{R}$
- (iii) $+$ on $\mathbb{Z}^-$
- (iv)   Scalar product on vectors in two dimensions

Solution

- (i)   (a) no   (b) yes

- (ii)   (a) yes   (b) no

[For example, $2 * (3 * 4) = |2 - |3 - 4|| = |2 - 1| = 1$, whilst

$(2 * 3) * 4 = ||2 - 3| - 4| = |1 - 4| = 3$]

(iii)	(a) yes	(b) yes	
(iv)	(a) yes	(b) no [as $(a * b) * c$ is undefined: $a * b$ is a scalar]	

If there is an element $e \in S$, such that $e * a = a * e = a$, for all $a \in S$, then e is said to be an **identity element** for the binary operation $*$.

If, for an element $a \in S$, there exists an element $b \in S$, such that $a * b = b * a = e$, then b is said to be an **inverse** of a (and a is an inverse of b).

Example 7.3

For each of the following binary operations, determine (a) an identity element, if it exists, and (b) if an identity exists, whether an inverse exists for every element.

(i)   Matrix multiplication on the set of 2×2 matrices

(ii) $x * y = |x - y|$, where S is $\mathbb{R}$

(iii) $+$ on $\mathbb{Z}^-$

(iv)   Scalar product on vectors in two dimensions

Solution

(i)   (a) $\begin{pmatrix} 1 & 0 \\ 0 & 1 \end{pmatrix}$

       (b) Inverses don't exist for matrices with a determinant of zero.

(ii)   (a) No identity element exists. [Note that, for $y < 0, 0 * y = |0 - y| = -y$, so that $e \neq 0$]

       (b) n/a

(iii)   (a) No identity element exists.

       (b) n/a

(iv)   (a) No identity element exists.

       (b) n/a

For small finite sets, a Cayley table (or combination table) is often the easiest way of defining a binary operation. It is simply a table showing $a * b$ for every combination of a and b in the set associated with the operation.

Example 7.4

Create the Cayley table for $\times_4$ on the set $\{0, 1, 2, 3\}$.

Solution

$\times_4$	0	1	2	3
0	0	0	0	0
1	0	1	2	3
2	0	2	0	2
3	0	3	2	1

Table 7.1

This is a two-way multiplication table. The values in the table are the remainder when each product is divided by 4.

Discussion point

→ How can commutativity of a binary operation be established from a Cayley table?

The Cayley table provides a convenient means of establishing identities and inverses. Thus, for the operation $\times_4$ on the set $\{0, 1, 2, 3\}$, the element 1 leaves each of the other elements unchanged, whether pre-multiplying or post-multiplying. Also, you see that the elements 0 and 2 do not possess an inverse, as no 1 appears in their rows (or columns).

The Cayley table also makes it clear whether the binary operation is closed.

Exercise 7.1

① Prove that each of the operations of $+_m$ and $\times_m$ on the set $\{0, 1, 2, \ldots, m-1\}$ is closed, commutative and associative.

② Suppose that f is a function with an inverse and that • is an associative binary operation. A new binary operation ◊ is defined by $x \lozenge y = f^{-1}(f(x) \bullet f(y))$.

Prove that ◊ is associative. Find $x \lozenge y$ in the following cases.

(i) $f(x) = x + 3$, $\bullet = \times$

(ii) $f(x) = \dfrac{2x-1}{x-1}$, $\bullet = +$

③ Write out

(i) the addition table for $\{0, 1, 2, 3, 4\}$ under $+_5$

(ii) the multiplication table for $\{0, 1, 2, 3, 4, 5, 6\}$ under $\times_7$.

④ The absolute difference between two numbers x and y is written as $x \sim y$, so that $x \sim y = |x - y|$.

(i) Make a Cayley table showing the results of applying the binary operation $\sim$ to the set $S = \{0, 1, 2, 3, 4\}$.

(ii) Is S closed under $\sim$?

(iii) Is the operation $\sim$ associative?

(iv) Is there an identity element? If so, what is it?

(v) Does each element have an inverse? If so, what are they?

⑤ A binary operation, ∘, is defined on the set $\mathbb{R}$ of real numbers by
$$a \circ b = \begin{cases} \text{maximum of } a \text{ and } b & \text{if } a \neq b \\ a & \text{if } a = b. \end{cases}$$

State whether ∘ is (i) closed and (ii) associative, giving your reasons.

Show that $\mathbb{R}$ does not contain an identity element for this operation.

Suggest a set S on which this operation can be defined which does contain an identity. Does every element of your set S have an inverse in S?

⑥ Answer these questions, and explain the connections with modular arithmetic.

(i) If today is Monday, what day of the week will it be after 100 days?

(ii) If a train departs at 21.37 and the journey takes 5 hours 40 minutes, at what time (on the 24-hour clock) does the train arrive?

(iii) Is 48 902 918 529 263 a perfect square?

⑦ Prove or disprove each of these statements:

(i) $x + 2 \equiv y + 4 \pmod{12}$
$\Rightarrow x \equiv y + 2 \pmod{12}$

(ii) $2x \equiv 4y \pmod{12} \Rightarrow x \equiv 2y \pmod{12}$.

⑧ Jay says that $\begin{pmatrix} 1 & -3 \\ 3 & 1 \end{pmatrix}$ is the inverse of $\begin{pmatrix} 1 & \frac{2}{3} \\ -\frac{2}{3} & 1 \end{pmatrix}$

under the binary operation

$A*B = AB - A - B$.

Determine whether she is correct.

⑨ The *union* of two sets A and B is the set $A \cup B$ of elements belonging to A or B or both. The *intersection* $A \cap B$ is the set of elements belonging to both A and B. Investigate whether the operations $\cup$ and $\cap$ are associative.

⑩ The operation $\circ$ is defined on the number pairs $A = (a_1, a_2)$ and $B = (b_1, b_2)$ so that $A \circ B = (a_1 + b_2, a_2 + b_1)$ and $A = B$ if and only if $a_1 = b_1$ and $a_2 = b_2$.
Find whether the operation is associative.

Find P such that $A \circ P = A$, and Q such that $Q \circ A = A$, and determine whether I, J can be found such that, for all A, $A \circ I = A$ and $J \circ A = A$.

Given that $A^* = (-a_2, -a_1)$, discuss the following:

$B \circ A = C \circ A \Rightarrow B \circ A \circ A^* = C \circ A \circ A^*$

$\Rightarrow B \circ (0, 0) = C \circ (0, 0)$

$\Rightarrow B = C$ [MEI]

2 The language of groups

A **group** $(G, *)$ has the following properties.

- G is a non-empty set.

- $*$ is a closed binary operation.

- $*$ is associative.

- There is an identity element, e.

- Each $a \in G$ has an inverse $a^{-1} \in G$.

Points to note

- The group $(G, *)$ is often abbreviated to just G, if the operation is understood.

- If the operation is commutative, the group is referred to as **abelian** (or just commutative). For non-abelian groups, the order of the elements in an operation will need to be specified.

- When there is no ambiguity, $a * b$ can be replaced by ab. However, when the operation is addition (including modular addition), $a + b$ is used instead of ab. By convention, if $+$ is used for the group operation, then the group is abelian; if ab is used, it may not be abelian.

- The order of the group G, which is sometimes denoted by $n(G)$, is the number of elements in the set G. The order may be infinite.

- When testing that the necessary conditions for a group apply, associativity is usually the time-consuming one, and is best saved until last (in case one of the other conditions fails).

- It can be shown that the identity element is unique, and that the inverse of each element is unique.

Example 7.5

Consider the set $\{0, 1, 2, 3, 4\}$ together with the operation of addition mod 5. Construct a Cayley table to establish whether this constitutes a group,

Solution

	0	1	2	3	4
0	0	1	2	3	4
1	1	2	3	4	0
2	2	3	4	0	1
3	3	4	0	1	2
4	4	0	1	2	3

Table 7.2

As the outcome of the binary operation on any two elements of the set is another element of the set, you can see that the operation is closed.

You see that the number 0 performs the role of the identity element.

Also, the identity element appears in each row, so that for each element a there is an element b such that $a + b = e$, and therefore each element has an inverse in the set.

It is possible to write $a +_5 b$, but $a + b$ is used when the modulus is understood.

Associativity of addition mod m follows from the associativity of ordinary addition, and hence the conditions for a group are satisfied. This group is commonly denoted $(\mathbb{Z}_5, +)$.

> **Note**
>
> In general, associativity needs to be proved, but it can just be asserted in the case of addition or multiplication (both ordinary and modular), and also for composition of functions.
>
> Note also that each element of the group appears once only in each row and column of the Cayley table (making the table a 'Latin square'). It can be shown that this is true for all groups.

Exercise 7.2

① Prove that if a and b are elements of a group, then $(ab)^{-1} = b^{-1}a^{-1}$.

② Prove that if $a^2 = e$ (the identity) for every element a of a group, then the group is abelian. Give an example of such a group.

③ Draw an appropriate Cayley table and hence determine whether the set $\{1, 2, 3, 4\}$ under multiplication (mod 5) forms a group.

④ Determine whether the set $\left\{ \begin{pmatrix} 1 & 0 \\ 0 & 1 \end{pmatrix}, \begin{pmatrix} 0 & 1 \\ -1 & 0 \end{pmatrix}, \begin{pmatrix} -1 & 0 \\ 0 & -1 \end{pmatrix}, \begin{pmatrix} 0 & -1 \\ 1 & 0 \end{pmatrix} \right\}$ under multiplication forms a group.

⑤ Determine whether the set $\{1, i, -1, -i\}$ under multiplication forms a group.

⑥ In the Cayley table of a group with identity e the following rectangles occur:

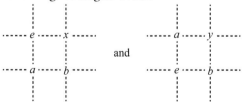

and

Figure 7.1

Express x and y in terms of a and b.

⑦ Which of the following sets of 2×2 matrices form a group under the operation of matrix multiplication? Which of these groups are abelian?

(i)   those with rational elements

(ii) those with integer elements and unit determinant

(iii) those of the form $\begin{pmatrix} \cos\theta & -\sin\theta \\ \sin\theta & \cos\theta \end{pmatrix}$

(iv) those of the form $\begin{pmatrix} z & -w^\star \\ w & z^\star \end{pmatrix}$ where z and w are complex numbers, not both zero.

3 Cyclic groups

The product $x * x * \ldots * x$, with x appearing n times, is abbreviated to x^n.

The **period** (or order) of an element x of a group is equal to the smallest non-negative integer n such that $x^n = e$. The period of e is 1.

For example, when the operation is addition, $x^3 = x + x + x$.

Example 7.6

What are the periods of the elements of $(\mathbb{Z}_5, +)$, other than e?

Solution

$e = 0$

$1^2 = 1 + 1 = 2; 1^3 = 1 + 1 + 1 = 3; 1^4 = 4; 1^5 = 5 \equiv 0 \pmod 5$; so the period of 1 is 5.

$2^2 = 2 + 2 = 4; 2^3 = 2 + 2 + 2 = 6 \equiv 1 \pmod 5; 2^4 = 8 \equiv 3 \pmod 5$; $2^5 = 10 \equiv 0 \pmod 5$; so the period of 2 is 5.

$3^2 = 6 \equiv 1 \pmod 5; 3^3 = 9 \equiv 4 \pmod 5; 3^4 = 12 \equiv 2 \pmod 5$; $3^5 = 15 \equiv 0 \pmod 5$; so the period of 3 is 5.

$4^2 = 8 \equiv 3 \pmod 5; 4^3 = 12 \equiv 2 \pmod 5; 4^4 = 16 \equiv 1 \pmod 5$; $4^5 = 20 \equiv 0 \pmod 5$; so the period of 4 is 5.

Thus every element of $(\mathbb{Z}_5, +)$ other than e has a period of 5.

It is said that g is a **generator** of the group G (of order n) if $g^n = e$, and no smaller power of g equals e. Then G can be written as $\{g, g^2, \ldots, g^n\}$, and G is said to be a **cyclic** group.

The group generated by g under the binary operation $*$ can be denoted by $\langle g, * \rangle$, or $\langle g \rangle$, when the binary operation is understood.

You have seen that every element of $(\mathbb{Z}_5, +)$, other than e, has a period of 5, and so it follows that every element other than e is a generator of $(\mathbb{Z}_5, +)$.

Cyclic groups can also be identified from the cyclic pattern of their Cayley tables. The table could be presented in any of the following forms.

	e	a	b	c
e	e	a	b	c
a	a	b	c	e
b	b	c	e	a
c	c	e	a	b

Table 7.3

	a	a^2	a^3	a^4
a	a^2	a^3	a^4	a
a^2	a^3	a^4	a	a^2
a^3	a^4	a	a^2	a^3
a^4	a	a^2	a^3	a^4

Table 7.4

	e	a	a^2	a^3
e	e	a	a^2	a^3
a	a	a^2	a^3	e
a^2	a^2	a^3	e	a
a^3	a^3	e	a	a^2

Table 7.5

Discussion point

➜ Are cyclic groups always abelian?

ACTIVITY 7.1
Show that the group {1, 2, 4, 8} under multiplication mod 15 is cyclic, by constructing a Cayley table.

Example 7.7

Find the generators of the group $\{1, i, -1, -i\}$ under multiplication.

Solution

The Cayley table is as follows.

	1	i	-1	-i
1	1	i	-1	-i
i	i	-1	-i	1
-1	-1	-i	1	i
-i	-1	1	i	-1

Table 7.6

$i^2 = -1, i^3 = -i, i^4 = 1$, so i is of period 4.

$(-1)^2 = 1$, so -1 is of period 2.

$(-i)^2 = -1, (-i)^3 = i, (-i)^4 = 1$, so $-i$ is of period 4.

Thus, i and $-i$ are generators of the group.

Exercise 7.3

① Prove that in a group the elements ab and ba have the same period.

② Create the Cayley table for the group $(\langle 4 \rangle, \times_7)$.

③ Prove that if an element a of a group has period n, then a^{-1} must have period n also.

④ (i) Show that C_5 (the cyclic group of order 5) can be generated by every one of its elements except for the identity. Show that the corresponding statement about C_6 is not true.

(ii) Investigate similar statements about C_7 and C_8.

(iii) Given that a is a generator of C_n, find a necessary and sufficient condition for a^k to generate C_n.

⑤ The Big Wheel at a fairground has 20 seats spaced equally around the rim. At the end of the ride the machinery automatically advances the wheel through successive angles of $\frac{\pi}{10}$ radians so that passengers may dismount in turn.

On one occasion, to the dismay of all the passengers except one, the machinery develops a fault at the end of the ride and begins to advance the wheel through successive angles of $\frac{n\pi}{10}$ where n is an integer. The remaining passenger, a mathematician, notices that $n = 9$ and patiently awaits his turn to dismount. Explain his confidence. What other values of n would leave him unperturbed?

[MEI]

⑥ The elements of the set $P = \{1, 3, 9, 11\}$ are combined under the binary operation $\star$, defined as multiplication modulo 16.

(i) Demonstrate associativity for the elements 3, 9, 11, in that order.

(ii) Assuming that associativity holds in general, show that P forms a group under the binary operation $\star$.

(iii) Write down the period of each element.

(iv) Show that the group $(P, \star)$ is cyclic.

4 Non-cyclic groups

Is $\{1, 3, 5, 7\}$ under multiplication mod 8 a cyclic group?

The Cayley table for this group is shown in Table 7.7.

	1	3	5	7
1	1	3	5	7
3	3	1	7	5
5	5	7	1	3
7	7	5	3	1

> The entries in the table are the remainders when each product is divided by 8.

Table 7.7

This group is not cyclic. It can in fact be shown that any group of order 4 is either cyclic or has the above structure, which is called the **Klein 4-group** (or sometimes Klein group V, where the V stands for the German 'vier', meaning four).

> **Note**
> You learned how to multiply matrices in *AQA A level Further Mathematics Year 1*.

ACTIVITY 7.2

What type of group is the following set, under matrix multiplication?

$$\left\{ e = \begin{pmatrix} 1 & 0 \\ 0 & 1 \end{pmatrix}, a = \begin{pmatrix} 1 & 0 \\ 0 & -1 \end{pmatrix}, b = \begin{pmatrix} -1 & 0 \\ 0 & 1 \end{pmatrix}, c = \begin{pmatrix} -1 & 0 \\ 0 & -1 \end{pmatrix} \right\}$$

Symmetry groups

The **symmetry group** of an equilateral triangle, D_3, is defined as follows.

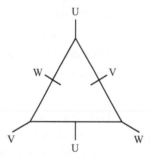

Figure 7.2

> **Note**
> D stands for 'dihedral' (meaning "two planes" – although the connection is rather obscure).
> More generally, D_n, the symmetry group of a regular n-sided polygon, is of order $2n$ (not n): in addition to the identity transformation, there will be $n-1$ rotations and n reflections.

There are 6 possible transformations of an equilateral triangle.

Let I be the identity transformation.

Let P be a rotation of 120° anticlockwise in the plane of the paper.

Let Q be a rotation of 120° clockwise in the plane of the paper.

Let U, V and W be reflections in the U, V and W axes, respectively.

Example 7.8

The Cayley table for D_3 is as follows.

	I	P	Q	U	V	W
I	I	P	Q	U	V	W
P	P	Q	I	W	U	V
Q	Q	I	P	V	W	U
U	U	V	W	I	P	Q
V	V	W	U	Q	I	P
W	W	U	V	P	Q	I

Table 7.8

> **Note**
>
> For example, the combined transformation $PU = W$ (from row P and column U) is to be interpreted as transformation U followed by transformation P (as for matrix multiplication).

(i) Is this group abelian?

(ii) What are the periods of each of the elements?

> **Solution**
>
> (i) No: for example, $UV = P$ but $VU = Q$.
>
> (ii) U, V and W are of period 2.
>
> $P^3 = P\,(P^2) = PQ = I$ and $Q^3 = Q\,(Q^2) = QP = I$, so that both P and Q are of period 3.
>
> Thus the elements I, P, Q, U, V, W have periods $1, 3, 3, 2, 2, 2$, respectively.

ACTIVITY 7.3

Consider the set $\left\{ x, 1-x, \dfrac{1}{x}, \dfrac{1}{1-x}, \dfrac{x-1}{x}, \dfrac{x}{x-1} \right\}$ under composition of functions on $x \in \mathbb{R}, x \neq 0, 1$

Let $e = x$, $a = 1-x$, $b = \dfrac{1}{x}$, $c = \dfrac{1}{1-x}$, $d = \dfrac{x-1}{x}$, $f = \dfrac{x}{x-1}$.

(i) Is this group abelian?

(ii) What are the periods of each of the elements?

> **Note**
>
> You will have met composite functions in GCSE Mathematics. Remember that gf(x) means g[f(x)].

① Write out the Cayley table for combining the symmetry transformations of a square, using I for the identity transformation, R, H, S for rotations 90° anticlockwise, 180° and 90° clockwise about O, respectively, and $X, Y, A, B,$ for reflections in axes $x, y, a, b,$ respectively.

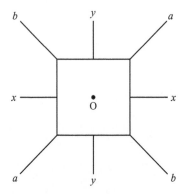

Figure 7.3

Check that all the group axioms are satisfied.

② (i) Show that the sets $\{1, 4, 7, 13\}$ and $\{3, 6, 9, 12\}$ form abelian groups under multiplication modulo 15, but $\{2, 5, 8, 14\}$ does not.

(ii) Write out the multiplication table for $\{1, 2, 3, 4, 5, 6, 7\}$ under multiplication modulo 8, and explain why this is not a group. Find subsets of this set which, under the same operation, give groups of order 2 and 4.

③ (i) Show that the set $\left\{x, -\dfrac{1}{x}, \dfrac{x-1}{x+1}, \dfrac{1+x}{1-x}\right\}$ forms a group under composition of functions on $x \in \mathbb{R}, x \neq -1, 0, 1$

(ii) Show that it is not the Klein 4-group.

(iii) Deduce the nature of the group.

④ The set $G = \{P, Q, R, S\}$ consists of the four matrices

$$P = \begin{pmatrix} 1 & 0 \\ 0 & 1 \end{pmatrix}, Q = \begin{pmatrix} -1 & 0 \\ 0 & -1 \end{pmatrix}, R = \begin{pmatrix} 0.6 & 0.8 \\ 0.8 & -0.6 \end{pmatrix},$$
$$S = \begin{pmatrix} -0.6 & -0.8 \\ -0.8 & 0.6 \end{pmatrix}.$$

(i) Show that G is a group under matrix multiplication. (You may assume that matrix multiplication is associative.)

(ii) State the period of each of the elements of G.

⑤ With the notation of question 1 of this Exercise for the group D_4:

(i) By expressing H, S, Y, A, B in terms of R and/or X, show that R and X are generators of the group.

(ii) Show that $R^4 = X^2 = I$ and $RX = XR^3$ (these are called *defining relations*).

⑥ Show that D_n is generated by two elements R and X with $R^n = X^2 = I$ and $RX = XR^{n-1}$.

5 Subgroups

Any subset of G that gives rise to a group under the operation $*$ is referred to as a **subgroup** of $\{G, *\}$.

Both $\{e\}$ and the whole group are subgroups. However, $\{e\}$ is referred to as the **trivial** subgroup, and a **proper** subgroup is any subgroup apart from the whole group (so that the **trivial** subgroup counts as a proper subgroup). A **non-trivial** subgroup is any subgroup other than $\{e\}$.

Any subgroup must satisfy the properties of a group. In particular, it must have an identity element, and this will be the identity element, e, of the parent group (as $ae = ea = a$ for any element of the subgroup).

> **Note**
>
> Some textbooks give different definitions of trivial subgroups and proper subgroups, but you are most likely to need the ones given on the right.

One way of finding subgroups of a group G is to examine the Cayley table. In addition to the identity element, you choose a set of elements that will ensure closure.

A more systematic way is to create the set $\{a, a^2, a^3, \dots\}$, where a is an element of G. This is referred to as the subgroup generated by a (it may, however, just be the parent group).

This applies to both finite and infinite G.

Example 7.9

Find any proper subgroups of the symmetry group of an equilateral triangle, D_3, in Example 7.8.

ACTIVITY 7.4

Find any proper subgroups of the group in Activity 7.3.

ACTIVITY 7.5

Find the proper subgroups of the Klein 4-group.

Solution

Investigate the subgroups generated by each element of G.

From the Cayley table (Table 7.8), $P^2 = Q$ and $P^3 = P.\,P^2 = P.\,Q = I$.

This gives the subgroup $\{I, P, Q\}$. This can also be noted directly from the Cayley table (helped by the fact that I, P and Q are next to each other in the table).

You can see that $U^2 = I$, $V^2 = I$ and $W^2 = I$, so that $\{I, U\}$, $\{I, V\}$ and $\{I, W\}$ are also subgroups. In addition, $\{I\}$ is a subgroup.

> Alternatively,
> $P^3 = P^2.\,P = PQ = I$;
> noting that transformations P and Q commute, even though the group as a whole is not abelian.

Lagrange's theorem

Discussion point

➔ How does Lagrange's theorem show that groups of a prime order are cyclic?

Lagrange's theorem states that the order of a subgroup of a finite group is a factor of the order of the group.

This is consistent with the subgroups you found in the previous section.

However, the converse of the theorem is not true: if the order of a group has a factor f, then there isn't necessarily a subgroup of order f.

By Lagrange's theorem, the period of an element divides the order of the group.

Exercise 7.5

① (i) Show that the group of symmetries of a square (D_4) has order 8.

(ii) Find the periods of all the elements of the group.

(iii) Find all the subgroups.

(iv) Identify any cyclic subgroups.

② Prove that if H and K are two subgroups of a group G, then $H \cap K$ is also a subgroup. Give an example to show that $H \cup K$ need not be a subgroup.

③ Find the subgroups of $(\mathbb{Z}_6, +)$.

④ G is a finite group which is commutative (i.e. $ab = ba$ for all $a, b \in G$). S and T are subgroups of G, and $S \cap T = \{e\}$, where e is the identity element of G.

Define ST to be the set of all elements of the form st, where $s \in S$ and $t \in T$.

(i) Show that ST is a subgroup of G.

(ii) Show that if $s_1 t_1 = s_2 t_2$ where $s_1, s_2 \in S$ and $t_1, t_2 \in T$, then $s_1 = s_2$ and $t_1 = t_2$.

(iii) Deduce that if S contains k elements and T contains m elements, then ST contains km elements.

Now let $G = \{1, 5, 7, 11, 13, 17, 19, 23, 25, 29, 31, 35\}$, where the binary operation is multiplication modulo 36. You may assume that G is a commutative group.

Three cyclic subgroups of G are

$S = \{1, 13, 25\}$, $T = \{1, 17\}$ and $U = \{1, 35\}$.

(iv) List the elements of the subgroup ST. Show that ST is cyclic.

(v) List the elements of the subgroup TU. Show that TU is not cyclic.

[MEI]

⑤ A non-abelian group G consists of eight 2×2 matrices, and the binary operation is matrix multiplication. The eight distinct elements of G can be written as

$G = \{I, A, A^2, A^3, B, AB, A^2B, A^3B\}$,

where I is the identity matrix, and A, B are 2×2 matrices such that

$A^4 = I,$ $\qquad B^2 = I,$ $\qquad$ and $\qquad BA = A^3B.$

(i) Show that $(A^2B)(AB) = A$ and $(AB)(A^2B) = A^3$.

(ii) Evaluate the following products, giving each one as an element of G as listed above: $(AB)(A)$, $(AB)(AB)$, $(B)(A^2)$.

(iii) Find the period of each element of G.

(iv) Show that $\{I, A^2, B, A^2B\}$ is a subgroup of G.

(v) Find the other two subgroups of G that have order 4.

(vi) For each of the three subgroups of order 4, state whether or not it is a cyclic subgroup.

[MEI]

⑥ Let $p = 2^k + 1$ be a prime number, and let G be the group of integers $1, 2, \dots, p - 1$, with multiplication defined modulo p.

(i) Show that if $0 < m < k$ then $0 < 2^m - 1 < p$ and deduce that $2^m \not\equiv 1 \pmod{p}$.

(ii) Show that if $k < m < 2k$ then $2^{2k-m} \not\equiv 1 \pmod{p}$ and deduce that $2^m \not\equiv 1 \pmod{p}$.

(iii) Use parts (i) and (ii) to show that the period of the element 2 in G is $2k$.

(iv) Deduce that k is a power of 2.

6 Isomorphisms

Two groups G and H are said to be **isomorphic** if they have the same structure, and you write $G \cong H$. This idea was met, in the context of graphs, in Chapter 1.

To prove that two finite groups are isomorphic, you need only verify that their Cayley tables have the same structures (though elements may need to be rearranged first).

Certain results can be shown to hold for any isomorphism, and an isomorphism can often be shown not to exist if these results do not hold.

For isomorphic groups G and H:

■ G is abelian if and only if H is abelian

■ the periods of the elements of G are the same as those of the elements of H

■ the orders of the subgroups of G are the same as those of H.

Example 7.10

Show that the groups $(\mathbb{Z}_4, +)$ and $\{1, i, -1, -i\}$ are isomorphic.

Solution

The Cayley tables for the two groups $(\mathbb{Z}_4, +)$ and $\{1, i, -1, -i\}$ are shown in Tables 7.9 and 7.10.

	0	1	2	3
0	0	1	2	3
1	1	2	3	0
2	2	3	0	1
3	3	0	1	2

Table 7.9

	1	i	−1	−i
1	1	i	−1	−i
i	i	−1	−i	1
−1	−1	−i	1	i
−i	−i	1	i	−1

Table 7.10

> **Note**
>
> When comparing Cayley tables, it may be necessary to change the order of the elements.

There is a one-to-one correspondence between the elements of the two groups, so the groups are isomorphic.

Other results for isomorphisms

- All cyclic groups of a particular order are isomorphic to each other, and to (for example) $(\mathbb{Z}_n, +)$.

- All groups of order 4 are isomorphic to either the cyclic group or the Klein 4-group.

- There are two distinct groups of order 6: cyclic groups and groups isomorphic to D_3.

Exercise 7.6

① Prove that it is impossible for an abelian group to be isomorphic to a non-abelian group.

② Prove that the group of positive real numbers under multiplication is isomorphic to the group of real numbers under addition. (Hint: Consider logarithms.)

③ Prove that the group of matrices $\begin{pmatrix} a & 0 \\ 0 & a \end{pmatrix}$ under matrix addition and the group of matrices $\begin{pmatrix} 1 & 0 \\ b & 1 \end{pmatrix}$ under matrix multiplication, where a and b are real numbers, are isomorphic.

④ Suppose that a group of order 4 has the distinct elements e, a, b, c, where e is the identity.

(i) Explain why ab cannot equal a or b.

(ii) Deduce that there are the following four possibilities.

(a) $ab = ba = e$ (b) $ab = e, ba = c$

(c) $ab = c, ba = e$ (d) $ab = ba = c$

(iii) By trying to complete the Cayley table, show that cases (b) and (c) lead to contradictions, that (a) gives the Cayley table for one group, and that (d) gives the Cayley tables for three groups.

⑤ Determine whether $(\mathbb{Z}_6, +)$ is isomorphic to D_3.

⑥ (i) Prove that all groups of order 2 are isomorphic.

(ii) Show that there is only one way of completing this Latin square.

$$
\begin{array}{ccc}
e & a & b \\
a & \dots & \dots \\
b & \dots & \dots
\end{array}
$$

Deduce that all groups of order 3 are isomorphic. Give a specific example of one such group.

⑦ Show that each of the following sets of four elements with the stated operation forms a group, and sort these six groups into isomorphic sets.

(i) 1, −1, i, −i; multiplication of complex numbers

(ii) functions

$$e(x) = x, f(x) = -\frac{1}{x}, g(x) = \frac{x-1}{x+1},$$

$$h(x) = \frac{1+x}{1-x}; \text{ composition of functions}$$

(iii) 1, 2, 3, 4; multiplication modulo 5

(iv) 1, 3, 5, 7; multiplication modulo 8

(v) $\begin{pmatrix} 1 & 0 \\ 0 & 1 \end{pmatrix}, \begin{pmatrix} 1 & 0 \\ 0 & -1 \end{pmatrix}, \begin{pmatrix} -1 & 0 \\ 0 & 1 \end{pmatrix}, \begin{pmatrix} -1 & 0 \\ 0 & -1 \end{pmatrix}$; matrix multiplication

(vi) $\begin{pmatrix} 1 & 0 \\ 0 & 1 \end{pmatrix}, \begin{pmatrix} 0 & 1 \\ -1 & 0 \end{pmatrix}, \begin{pmatrix} -1 & 0 \\ 0 & -1 \end{pmatrix}, \begin{pmatrix} 0 & -1 \\ 1 & 0 \end{pmatrix}$; matrix multiplication

⑧ The notation $\begin{bmatrix} 1 & 2 & 3 & 4 \\ 2 & 4 & 1 & 3 \end{bmatrix}$

denotes the permutation, or mapping of {1, 2, 3, 4} onto itself, in which 1 has image 2, 2 has image 4, 3 has image 1, 4 has image 3. Permutations are combined in the usual way by composition of mappings.

Show that the set of permutations

$$\left\{ \begin{bmatrix} 1 & 2 & 3 & 4 \\ 1 & 2 & 3 & 4 \end{bmatrix}, \begin{bmatrix} 1 & 2 & 3 & 4 \\ 2 & 1 & 4 & 3 \end{bmatrix}, \begin{bmatrix} 1 & 2 & 3 & 4 \\ 3 & 4 & 1 & 2 \end{bmatrix}, \begin{bmatrix} 1 & 2 & 3 & 4 \\ 4 & 3 & 2 & 1 \end{bmatrix} \right\}$$

and the set of matrices

$$\left\{ \begin{pmatrix} 1 & 0 \\ 0 & 1 \end{pmatrix}, \begin{pmatrix} 1 & 0 \\ 0 & -1 \end{pmatrix}, \begin{pmatrix} -1 & 0 \\ 0 & 1 \end{pmatrix}, \begin{pmatrix} -1 & 0 \\ 0 & -1 \end{pmatrix} \right\}$$

under matrix multiplication are isomorphic groups, and that the set of complex numbers {1, −1, i, −i} (under multiplication) is a group not isomorphic with the others.

For each group give the period of each element, say whether the group is cyclic, and list all its proper subgroups.

Indicate briefly how you would show that these are the only two types of group of order 4.

[MEI]

⑨ The set G = {1, 3, 7, 9, 11, 13, 17, 19} is a group under the binary operation of multiplication modulo 20.

(i) Give the combination table for G.

(ii) State the inverse of each element of G.

(iii) Find the period of each element of G.

(iv) List all the subgroups of G.

Identify those subgroups which are isomorphic to one another.

(v) For each of the following, state, giving reasons, whether or not the given set and binary operation is a group. If it is a group, state, giving a reason, whether or not it is isomorphic to G.

(a) J = {0, 1, 2, 3, 4, 5, 6, 7} under multiplication modulo 8

(b) K = {0, 1, 2, 3, 4, 5, 6, 7} under addition modulo 8

[MEI]

LEARNING OUTCOMES

Now you have finished this chapter, you should be able to

➤ understand and use binary operations; including use of modular arithmetic and matrix multiplication

➤ understand, use and prove the commutativity of a binary operation

➤ understand, use and prove the associativity of a binary operation

➤ construct a Cayley table for a given set and binary operation

➤ understand and prove the existence of an identity element for a given set under a given binary operation

- ➤ find the inverse of an element belonging to a given set under a given binary operation
- ➤ understand and use the language of groups; including order, period, subgroup, proper, trivial, non-trivial
- ➤ understand and use the group axioms: closure, identity, inverses and associativity; including use of Cayley tables
- ➤ recognise and use finite and infinite groups and their subgroups; including groups of symmetries of regular polygons, cyclic groups and abelian groups
- ➤ understand and use Lagrange's theorem
- ➤ identify and use the generators of a group
- ➤ recognise and find isomorphism between groups of finite order.

KEY POINTS

1 A binary operation ∗ on a set S is a rule that assigns the element $x * y$ to any ordered pair of elements x, y in S. Binary operations may or may not be closed, commutative or associative.

2 The set may contain an identity element and elements of the set may have an inverse.

3 The integers x and y are said to be congruent mod m if the difference between them is a multiple of m.

4 A binary operation may be represented in a Cayley table.

5 A group is defined for a closed, associative binary operation, where an identity element exists and each element has an inverse.

6 If the operation is commutative, the group is called abelian.

7 The period of an element x of a group is the smallest non-negative integer n such that $x^n = e$.

8 G is a cyclic group if $g^n = e$ for some generator g.

9 The symmetry group of an n-sided polygon has order $2n$.

10 Any subset of G that gives rise to a group under the operation ∗ is referred to as a subgroup of $\{G, *\}$.

11 Lagrange's theorem states that the order of a subgroup of a finite group is a factor of the order of the group.

12 Two groups are isomorphic if they have the same structure.

13 For isomorphic groups G and H
- G is abelian if and only if H is abelian
- the periods of the elements of G are the same as those of the elements of H
- the orders of the subgroups of G are the same as those of H

14 Results for isomorphisms:
- all cyclic groups of a particular order are isomorphic to each other
- all groups of order 4 are isomorphic to either the cyclic group or the Klein 4-group
- there are two distinct groups of order 6: cyclic groups and groups isomorphic to D_3.

Chapter 1

Opening activity Page 1

Yes

Discussion point Page 3

The smallest number of edges occurs when the graph is a tree, when the number of edges will be $n - 1$.

Activity 1.1 Page 4

Every edge contributes two to the sum of the degrees of the vertices, so that this sum must be even. As the sum of the degrees of the even vertices is even, the sum of the degrees of the odd vertices must also be even, in order to give an even sum overall. So the number of odd vertices must be even.

Exercise 1.1 Page 4

1   (i)   ABCA, ABCDEA, ACEA, ACDEA, ABCEA
    (ii) ABCEDCA is not a cycle since vertex C is repeated.
    (iii) A (closed) trail. It is a walk too, but since no edge is repeated it is best described as a trail.

2   Possible examples are

3   (i)

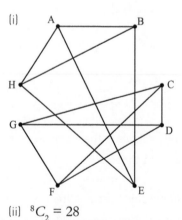

    (ii) $^8C_2 = 28$

4

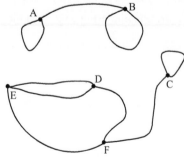

5

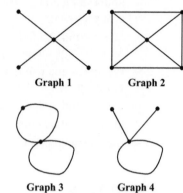

| Graph 1 | Graph 2 |

| Graph 3 | Graph 4 |

6   (i) $14 = 2 \times 7$, because each of the 7 edges has 2 ends.
    (ii) Two of the following:

	Number of degree 2	Number of degree 3	Number of degree 4
(a)	3	0	2
(b)	2	2	1
(c)	1	4	0

    (iii) Possible examples for each of the possibilities:

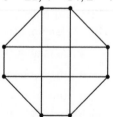

    (a)   (b)   (c)

7   (i) $6 - 21; 4 - 14; 2 - 7$

    (ii) $d = 2, 4, 5, 6$ or 7
    (iii) The number of edges $= \frac{1}{2} \times$ no. of nodes $\times$ degree of the nodes. Since the number of nodes is odd, the degree must be even to have an integer value for the number of edges. When there are 8 vertices, the number of edges will be an integer, whether d is odd

Activity 1.2 Page 6
Penzance

Activity 1.3 Page 6
$F + V = 10 + 11 = 21$

$E + 2 = 19 + 2 = 21$

Exercise 1.2 Page 8
1 Graph 1 – (ii) and (iii)
Graph 2 – (i)
Graph 3 – (iii)
Graph 4 – (ii) and (iii)

2 Graph 1, $V = 8, F = 4, E = 10; F + V = 8 + 4 = 12$;
$E + 2 = 10 + 2 = 12$
Graph 2, $V = 3, F = 3, E = 4; F + V = 3 + 3 = 6$;
$E + 2 = 4 + 2 = 6$
Graph 3, $V = 4, F = 5, E = 7; F + V = 4 + 5 = 9$;
$E + 2 = 7 + 2 = 9$
Graph 4, $V = 5, F = 4, E = 7; F + V = 5 + 4 = 9$;
$E + 2 = 7 + 2 = 9$

3 For example,

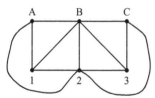

$V = 6, F = 7, E = 11, 6 + 7 = 11 + 2$

4 (i)

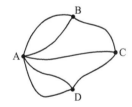

(ii) The four vertices of the graph are all odd, so that the graph is neither Eulerian nor semi-Eulerian.

5 (i)

(ii)

(iii) Three connection points and three internal points

6 (i)

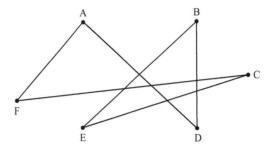

(ii)

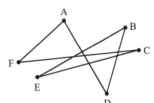

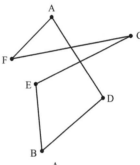

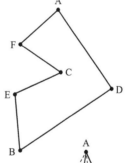

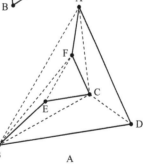

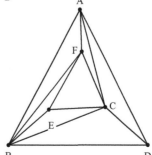

Discussion point Page 11

No. You cannot obtain K_4 from Figure 1.25 by stripping out edges and isolated vertices. You are not allowed to just remove vertex E, as that would leave the edges AE and ED with a vertex missing.

Discussion point Page 12

mn

Discussion point Page 12

Because the vertices on one side are not connected to each other.

Exercise 1.3 Page 12

1

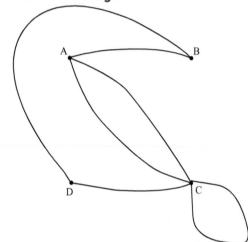

2

	A	B	C	D	E
A	2	0	0	0	0
B	0	0	1	1	1
C	0	1	0	0	0
D	0	1	0	0	1
E	0	1	0	1	0

3

	A	B	C	D	E
A	0	1	0	1	0
B	1	0	1	1	1
C	0	1	0	0	0
D	1	1	0	0	0
E	0	1	0	0	0

4 (i) It is Eulerian when it has an odd number of vertices, as that makes each vertex even. It can only be semi-Eulerian when there are two odd vertices.

(ii) It is Eulerian when each set has an even number of vertices; it is semi-Eulerian when one set has two vertices and the other has an odd number of vertices, or if each set has one vertex.

5 For example, all the possible edges are drawn in. To draw an edge you choose two vertices to join. So the number of edges is $^nC_2 = \frac{1}{2}n(n-1)$

Or: Each of the n vertices has $n-1$ edges leading from it, giving a total of $n(n-1)$ edges. However, each of the edges is being counted twice (once for each of the vertices at its ends), and so you divide by 2 to remove the duplication, giving $\frac{1}{2}n(n-1)$ edges.

6 (i)

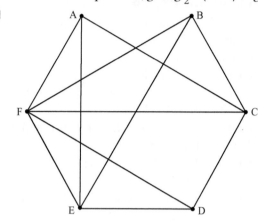

(ii)

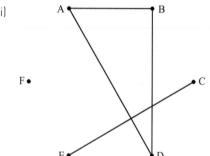

Schedule A, B and D together and schedule C and E together.

(iii) The duration of the tasks

7

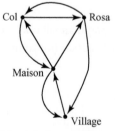

The lift from Village to Maison

8 (i) AB; ACB; ADB; ACDB; ADCB

(ii) 1 direct, AB

3 through one other point

$^3P_2 = 6$ through 2 other points

$3! = 6$ through 3 other points

$1 + 3 + 6 + 6 = 16$

Exercise 1.4 Page 14

1   Label the first matrix A, B, C, D. Label the second matrix C, A, D, B and it can clearly be seen that they are isomorphic.

2   (2) and (6); (1), (3), (4) and (5)

3   (i)   For example A↔1; D↔2; E↔3; B↔5; C↔6; F↔4

    (ii)

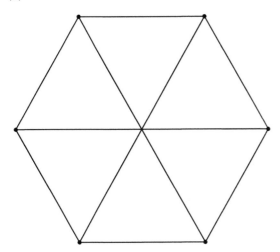

4   B is a 4-node and connects to C which is a 2-node in the first graph. The only 2-node in the second graph, U, does not connect with the only 4-node, which is Q and so they cannot be isomorphic.

5   P – B; Q – E; R – A; S – C; T – D

Activity 1.4 Page 17

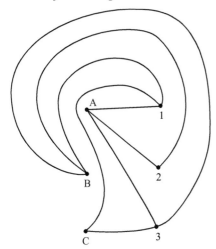

Exercise 1.5 Page 19

1   Graph 1 is planar as it does not contain a subdivision K_5
    Graph 2 is planar as it does not have a subgraph that is K_5 or $K_{3,3}$.

Graph 3 is not planar as K_5 is a subgraph of it (remove B or D)
Graph 4 is not planar as it has $K_{3,3}$ as a subgraph (lose C and 4)

2   K_5 has 10 edges so a graph with 8 edges does not have a subgraph that is K_5. $K_{3,3}$ has 9 edges so a graph with 8 edges does not contain a subgraph that is $K_{3,3}$. Hence result.

3   (i) $K_{4,2}$ does not have a subgraph that is $K_{3,3}$, as there are only two vertices in the second set of vertices.

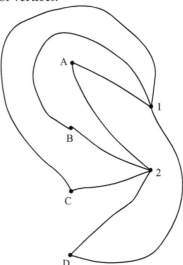

    (ii) $K_{n,2}$ does not have a subgraph that is $K_{3,3}$, as there are only two vertices in the second set of vertices.

4   (i)   No   (ii) No   (iii) Yes
    (iv) No   (v) Yes

5   Yes (The graph doesn't contain a subgraph that is a subdivision of K_5.)

6   (i)   (a) a–c1; b–c2; c–c3; d–c1; e–c4
          (b) a–c1; b–c2; c–c3; d–c2; e–c1
    (ii) (a) The vertex would have to be coloured differently to itself!
          (b) One edge is enough to force a different colour. A second adds nothing.
    (iii) (a)

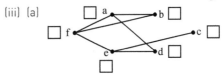

          (b) a–c1; b–c2; c–c1; d–c2; e–c3; f–c4
          (c) colour = hour;
               a & c then b & d then e then f
          (d) e.g. a & c then b & e then d & f

7   Draw a complete graph on six vertices. Imagine colouring the edge connecting two vertices green if the people represented by the vertices know

each other, and red if they do not. Choose a vertex. It has 5 edges incident upon it, so there must be three of the same colour, c1 say. Now examine the three edges joining vertices at the other ends of those three. If they are all coloured c2, then there is a c2 triangle. If not, then at least one is coloured c1, and that means that there is a c1 triangle which includes the original vertex.

8 Every person must map to a number. If there are n people then the image set is $\{0, 1, 2, ..., n-1\}$. But both 0 and $n-1$ cannot both be images since it is not possible simultaneously to have had somebody shake hands with everybody, and somebody else shake hands with nobody. Thus there are $n-1$ possible images for n subjects, so the mapping is many–one.

Chapter 2

Opening activity Page 22
For example, ABCDCBA: 14

Exercise 2.1 Page 26

1 Bod–Tru 26; Tru–Pen 27; Bod–Bud 30; Bud–Oke 30; Oke–Exe 23; Exe–Tor 22; Bod–Ply 31 (or could choose any other of the 31 weight arcs from Plymouth); Oke–Bar 31; Exe–Tau 35; Tau–Min 24.

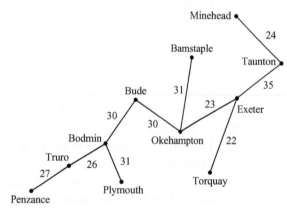

Total weight 279

2

	A	B	C	D	E	F	G	H
A	–	4	–	2	6	–	–	–
B	4	–	1	–	–	–	–	–
C	–	1	–	5	–	–	–	2
D	2	–	5	–	3	–	–	8
E	6	–	–	3	–	2	5	–
F	–	–	–	–	2	–	2	–
G	–	–	–	–	5	2	–	17
H	–	–	2	8	–	–	17	–

3

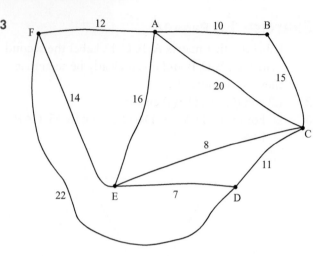

4 AB 10
AF 12
FE 14
ED 7
EC 8
Total weight = 51

5 ED 7
EC 8
AB 10
CD 11 (exclude)
AF 12
FE 14

6 (i) 107 miles

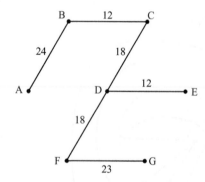

(ii) Add HA and HB and remove AB; 106 miles

7 (i)

From\to	A	B	C	D	E	M	S
A	–	–	–	–	–	–	50
B	60	–	–	–	–	50	$\sqrt{9700}$
C	$\sqrt{9700}$	50	–	50	$\sqrt{9700}$	60	–
D	–	–	–	–	60	50	$\sqrt{9700}$
E	–	–	–	–	–	–	50
M	50	–	–	–	50	–	60
S	–	–	–	–	–	–	–

(ii) e.g. ES; AS; MA; DM; BM; CB: total length = 300 m; drains correctly, since there is no house at which water accumulates.

(iii) Not worthwhile; can use ES; AS; DE; CD; BA with total length = 270 m.

Activity 2.1 Page 29

A possible route is ABCBFBDFEDCA.

In order to ensure that the arcs CB and BF are repeated, a useful device is to repeat each one straightaway: thus CB comes after BC and FB comes after BF.

Activity 2.2 Page 30

A possible route is: ACABFEFDBCDEA (which repeats AC and EF).

Activity 2.3 Page 30

A is an odd node, so you are happy to start there, if you don't need to finish there. B is an even node, so if you are to finish there then you need to turn it into an odd node by repeating a suitable path. Thus, the problem is the same as for category C(i), but with the four nodes to be dealt with being B, C, E and F (instead of A, C, E and F).

Activity 2.4 Page 30

The relevant pairings are

BC	4
BE	14 (BDE)
BF	7
CE	15 (CDE)
CF	11 (CBF)
EF	10

The possible ways of pairing up the nodes are

BC EF	4 + 10 = 14
BE CF	14 + 11 = 25
BF CE	7 + 15 = 22

The combination that gives the shortest total distance is thus BC EF, and the shortest route becomes 74 + 14 = 88.

Discussion point Page 30

You can select two of the odd nodes to be the start and end points, so that the path between them no longer has to be repeated. As AC is the shortest path between any pair of odd nodes, this is the one that you want to

repeat, and so you choose E and F as the start and end nodes (or the other way round).

Exercise 2.2 Page 32

1 (i) e.g. 1, 3, 1, 2, 3, 4, 1 (length = 87)

(ii) e.g. 1, 2, 3, 4, 1, 3 (length = 76)

2 For example, OYCYBLBOLCO, 1190 miles

3 For example, OYCYBLBOLCOL, 1247 miles

4 For example, YBOYCOLCLB, 979 miles

5 For example, Okehampton, Barnstaple, Bude, Okehampton, Bodmin, Truro, Plymouth, Exeter, Torquay, Plymouth, Bodmin, Bude, Barnstaple, Minehead, Taunton, Barnstaple, Exeter, Taunton, Exeter, Okehampton. 655 miles.

6 For example, Taunton, Exeter, Okehampton, Barnstaple, Bude, Okehampton, Bodmin, Truro, Plymouth, Exeter, Torquay, Plymouth, Bodmin, Bude, Barnstaple, Minehead, Taunton, Barnstaple, Exeter. 620 miles.

7 (i) 83 + 18

(ii) e.g. $D \rightarrow I \rightarrow H \rightarrow F \rightarrow D \rightarrow H \rightarrow G \rightarrow F \rightarrow$
$C \rightarrow B \rightarrow F \rightarrow C \rightarrow E \rightarrow G \rightarrow H \rightarrow I \rightarrow$
$A \rightarrow B \rightarrow A \rightarrow D$

(iii) 3 times

Discussion point Page 36

Prim's algorithm joins the nearest new node to **any** existing node, whereas the nearest neighbour algorithm joins it to the last node obtained. Also, Prim's algorithm is designed to produce a tree and you don't return to the start node.

Activity 2.5 Page 37

Arcs are added in the following order.

Tau–Min 24
Min–Bar 38
Bar–Oke 31
Oke–Exe 23
Exe–Tor 22
Tor–Ply 31
Ply–Bod 31
Bod–Tru 26

Unfortunately, you are now stranded, and so the algorithm breaks down for this particular starting point. The algorithm is completed by taking each of the other nodes to be the starting point, in turn.

Exercise 2.3 Page 38

1 Remove A, 21; remove B, 21; remove C, 24; remove D, 22; remove E, 33; remove F, 21; remove G, 21; remove H, 24. The lower bound is 33.

2 Starting with A, 33; starting with B, can't return; starting with C, 33; starting with D, 33; starting with E, 33; starting with F, 33; starting with G, 33; starting with H, 33. The upper bound is 33.

3 Remove A, 66; remove B, 66; remove C, 62; remove D, 62; remove E, 63; remove F, 66. The lower bound is 66.

4 Starting with A, 74; starting with B, 69; starting with C, 74; starting with D, 74; starting with E, 69; starting with F, 74. The upper bound is 69.

5

	A	B	C	D	E	F
A	–	3	7	13	6	5
B	3	–	4	10	3	2
C	7	4	–	6	5	6
D	13	10	6	–	11	12
E	6	3	5	11	–	1
F	5	2	6	12	1	–

6 (i) Weston → Burnham → Bridgewater → Glastonbury → Wells → Bath → Cheddar → Weston; 95

(ii) Weston → Burnham → Bridgewater → Glastonbury → Wells → Cheddar → Bath → Weston; 103

7 (i) Birmingham → Gloucester → Hereford → Shrewsbury → Stoke → Sheffield → Nottingham → Northampton → Birmingham; 357

(ii) The roads may be slower so the time is longer than an alternative. It may not provide appropriate stopping or refuelling points.

Exercise 2.4 Page 39

1 (i) 3

(ii) $5 \times 3 = 15$

(iii) $7 \times 5 \times 3 = 105$

(iv) 654729075

2 (i) B, D, F, G

(ii) Eulerian graphs have no odd nodes.

(iii) BD/FG = 550, BF/DG = 250, BG/DF = 550
e.g. A → B → F → B → C → D → E → G → E → D → G → F → E → C → A, Length = 1950 + 250 = 2200

3 (i) Strawberry → orange → lemon → lime → raspberry → strawberry; 73 minutes

(ii) No; this gives 77 minutes

4 A_1 can be added to the table with distance to and from A of ∞ and with distances to and from B, C, D and E equal to the distances from A to the towns. The solution to the travelling salesperson problem will then be of the form A ★★★A_1★★★A, and this can be interpreted as two separate Hamilton cycles, one for each lorry. The two tours are A → B → C → A and A → D → E → A; total length 237

5 (i) (a) All of order 2

(b) All nodes are even so the network is Eulerian.

Single rope needs only to be as long as the sum of the four section lengths.

(ii) (a)

Node	A	B	C	D	E	F
Order	3	3	1	1	2	2

In traversing a network vertices are encountered and left along different arcs. So there needs to be an even number of arcs incident upon each node.

(b) Some sections will have to have rope along them twice, so the total length required will exceed the sum of the lengths of the six sections.

(c) Need to have AD and BC repeated, e.g. C → B → F → A → D → A → E → B → C

(d) C → B → F → A → D → A → E → B or C → B → E → A → D → A → F → B or C → B → X → A → Y → B → Z → A → D, where {X, Y, Z} is any permutation of {E, E, F} or {E, F, F}.
(There are six such possibilities.)

6 (i) A → B → D → E → C → F → A with cost 640
Another tour is A → B → D → E → F → C → A with cost 537

(ii) $5! = 120$

(iii)

From \ To	A	B	C	D	E	F
A	–	65	80	78	110	165
B	75	–	97	55	113	130
C	80	90	–	70	90	340
D	90	65	90	–	75	250
E	110	90	80	45	–	82
F	165	130	320	195	100	–

(iv) e.g. B to E. Without taxes cheapest is BDE, costing 60. With taxes cheapest is BE costing 113.

(v) No difference. All taxes are incurred once and only once giving an increase of 140 on all tours.

7 (i) The colour is equivalent to the town, the cost of cleaning to the cost of travel, and using a colour to visiting a town.

(ii) WYGBRW – 13
YGBRWY – 13
BGYRWB – 17 or BGYWRB – 13 or BGRYWB – 14
GBRYWG – 12
RBGYWR – 13 or RGBYWR – 15

(iii) GBRYWG = WGBRYW

(iv) The two lowest weight arcs leading from the deleted node have to be an incoming arc and an outgoing arc, and this cannot be guaranteed in the case of a digraph.

Chapter 3

Opening activity Page 43

Cost of materials, How much people are prepared to pay, Number of potential customers, etc.

Activity 3.1 Page 46

At A, $3x + 4y = 24$ and $y = 3x \Rightarrow 15x = 24 \Rightarrow x = \frac{8}{5}$

so that A is (1.6, 4.8), with $P = 8$

Exercise 3.1 Page 46

1 $x = 3\frac{1}{7}, y = 2\frac{6}{7}, P = 8\frac{6}{7}$

2 $x = 4\frac{1}{4}, y = 3\frac{1}{6}, P = 55\frac{1}{12}$

3 Let x be the number of minutes spent walking.
Let y be the number of minutes spent running.
Maximise $D = 90x + 240y$
subject to $90x + 720y \leqslant 9000$
$$x + y \leqslant 30$$
Answer: $x = 20, y = 10, D = 4200$

4 13 luxury and 17 standard, giving a profit of £335 000

5 $x = 1; y = 2; z = 3$

6 (i) $x + y \leqslant 4$
$x \geqslant 2$
$y \geqslant 0.6$
$y \geqslant 0.25x$

(ii)

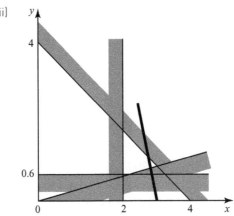

(iii) £3.2 million

(iv) 2.3

(v) See heavy line on graph in part (ii). Points of equal (fan) satisfaction

(vi) £3 382 000 in total; £618 000 less £2 706 000 on the playing squad; £206 000 more

7 (i) Let x be the number of units of X produced and y be the number of units of Y produced.
Maximise $x + y$
subject to $15x + 5y \leqslant 600$
$$10x + 7y \leqslant 560$$
$$8x + 12y \leqslant 768$$

(ii)

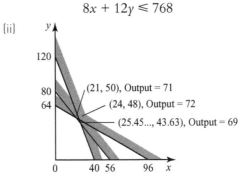

(21, 50), Output = 71
(24, 48), Output = 72
(25.45..., 43.63), Output = 69

(iii) The second constraint becomes
$10x + 7y \leq 576$, making (24, 48) feasible.

(iv) Any further increase in availability of B is
irrelevant since the constraint is now not
active.

8 Let x be the number of tonnes of deep-mined
and y the number of tonnes of opencast.
Maximise $10x + 15y$ (or minimise $10x + 5y$)
subject to $x + y = 20\,000$
$2x + y \leq 34\,000$ (chlorine)
$3x + y \leq 40\,000$ (sulphur)
$35x + 10y \leq 400\,000$ (ash)
$5x + 12y \leq 200\,000$ (water)
Gives $x \approx 5700$; $y \approx 14\,300$, with water constraint
critical and the others redundant.

9 12 tables and 30 chairs; profit = £390.
It is likely that the demand for chairs will be
greater than the supply.

10 (i) Let x be the number of the cheaper jacket
and y be the number of the more expensive
jacket.
Maximise $P = 10x + 20y$
subject to $x + y \geq 200$
$10x + 30y \leq 2700$
$20x + 10y \leq 4000$.

(ii) $x = 186$; $y = 28$; $P = 2420$.

(iii) The purchase constraint exceeds the
maximum order allowed by the other
constraints $(220 > 214)$.

Exercise 3.2 Page 53

1

P	x	y	z	s_1	s_2	Value
1	−9	−10	−6	0	0	0
0	2	3	4	1	0	3
0	6	6	2	0	1	8
1	$-\frac{7}{3}$	0	$\frac{22}{3}$	$\frac{10}{3}$	0	10
0	$\frac{2}{3}$	1	$\frac{4}{3}$	$\frac{1}{3}$	0	1
0	2	0	−6	−2	1	2
1	0	0	$\frac{1}{3}$	1	$\frac{7}{6}$	$\frac{37}{3}$
0	0	1	$\frac{10}{3}$	1	$-\frac{1}{3}$	$\frac{1}{3}$
0	1	0	−3	−1	$\frac{1}{2}$	1

Solution: $x = 1, y = \frac{1}{3}, z = 0, P = 12\frac{1}{3}$

2

P	w	x	y	z	s_1	s_2	Value
1	−3	−2	0	0	0	0	0
0	1	1	1	1	1	0	150
0	2	1	3	4	0	1	200
1	0	−0.5	4.5	6	0	1.5	300
0	0	0.5	−0.5	−1	1	−0.5	50
0	1	0.5	1.5	2	0	0.5	100
1	0	0	4	5	1	1	350
0	0	1	−1	−2	2	−1	100
0	1	0	2	3	−1	1	50

Solution: $w = 50, x = 100, y = 0, z = 0, P = 350$

3

P	w	x	y	z	s_1	s_2	s_3	Value
1	−3	−2	0	0	0	0	0	0
0	1	1	1	1	1	0	0	150
0	2	1	3	4	0	1	0	200
0	−1	1	0	0	0	0	1	0
1	0	−0.5	4.5	6	0	1.5	0	300
0	0	0.5	−0.5	−1	1	−0.5	0	50
0	1	0.5	1.5	2	0	0.5	0	100
0	0	1.5	1.5	2	0	0.5	1	100
1	0	0	5	$\frac{20}{3}$	0	$\frac{5}{3}$	$\frac{1}{3}$	$\frac{1000}{3}$
0	0	0	−1	$-\frac{5}{3}$	1	$-\frac{2}{3}$	$-\frac{1}{3}$	$\frac{50}{3}$
0	1	0	1	$\frac{4}{3}$	0	$\frac{1}{3}$	$-\frac{1}{3}$	$\frac{200}{3}$
0	0	1	1	$\frac{4}{3}$	0	$\frac{1}{3}$	$\frac{2}{3}$	$\frac{200}{3}$

Solution: $w = 66\frac{2}{3}, x = 66\frac{2}{3}, y = 0, z = 0, P = 333\frac{1}{3}$

4 (i) a is the number of aardvarks, etc.
First inequality models the furry material
constraint
Second inequality models the woolly material
constraint
Third inequality models the glass eyes constraint
That would model a 'pairs of glass eyes'
constraint

(ii) The problem is an IP, so the number of eyes
used will be integer anyway

(iii) e.g.

P	a	b	c	s_1	s_2	s_3	Value
1	-3	-5	-2	0	0	0	0
0	0.5	1	1	1	0	0	11
0	2	1.5	1	0	1	0	24
0	2	2	2	0	0	1	30
1	-0.5	0	3	5	0	0	55
0	0.5	1	1	1	0	0	11
0	1.25	0	-0.5	-1.5	1	0	7.5
0	1	0	0	-2	0	1	8
1	0	0	2.8	4.4	0.4	0	58
0	0	1	1.2	1.6	-0.4	0	8
0	1	0	-0.4	-1.2	0.8	0	6
0	0	0	0.4	-0.8	-0.8	1	2

Make 6 aardvarks and 8 bears giving £58 profit. 2 eyes are left over.

5 (i) Let x be the number of maths books produced.
Line 1 $\Leftrightarrow$ max $6x + 3y + 7z$ ($10 - 4 = 6$, etc.)
Line 2 $\Leftrightarrow$ $2x + 1.5y + 2.5z \leqslant 10\,000$
(printing time)
Line 3 $\Leftrightarrow$ $x + 0.5y + 1.5z \leqslant 7500$
(packing time)
Line 4 $\Leftrightarrow$ $300x + 200y + 400z \leqslant 2\,000\,000$
(storage space)

(ii)

P	x	y	z	s_1	s_2	s_3	Value
1	-6	-3	-7	0	0	0	0
0	2	1.5	2.5	1	0	0	10000
0	1	0.5	1.5	0	1	0	7500
0	300	200	400	0	0	1	2000000
1	-0.4	1.2	0	2.8	0	0	28000
0	0.8	0.6	1	0.4	0	0	4000
0	-0.2	-0.4	0	-0.6	1	0	1500
0	-20	-40	0	-160	0	1	400000
1	0	1.5	0.5	3	0	0	30000
0	1	0.75	1.25	0.5	0	0	5000
0	0	-0.25	0.25	-0.5	1	0	2500
0	0	-25	25	-150	0	1	500000

Produce 5000 maths books –
at a profit of £30 000
(2500 packing minutes spare and 0.5 m³ storage space spare.)

(iii) £1.50 and 50p respectively.

Exercise 3.3 Page 54

1 (i) Let a be the number of tonnes of Product A, etc.
Max $a + b + c$
st $3a + 2b + 5c < 60$
$5a + 6b + 2c < 50$

(ii) e.g.

P	a	b	c	s_1	s_2	Value
1	-1	-1	-1	0	0	0
0	3	2	5	1	0	60
0	5	6	2	0	1	50
1	-0.4	-0.6	0	0.2	0	12
0	0.6	0.4	1	0.2	0	12
0	3.8	5.2	0	-0.4	1	26
1	>0	0	0	>0	>0	15
0		0	1			10
0	$\frac{19}{26}$	1	0	$-\frac{2}{26}$	$\frac{5}{26}$	5

Make 5 tonnes of B and 10 tonnes of C

2 (i) a is the number of acres of land put to crop A, etc
$a + b \leqslant 20$ is equivalent to $a + b \leqslant c + d$
Given that $a + b + c + d \leqslant 40$, the maximisation will ensure that
$a + b + c + d = 40$ (and it's easier to solve using Simplex).

(ii)

P	a	b	c	d	s_1	s_2	Value
1	-50	-40	-40	-30	0	0	0
0	1	1	0	0	1	0	20
0	1	1	1	1	0	1	40
1	0	10	-40	-30	50	0	1000
0	1	1	0	0	1	0	20
0	0	0	1	1	-1	1	20
1	0	10	0	10	10	40	1800
0	1	1	0	0	1	0	20
0	0	0	1	1	-1	1	20

20 acres to A and 20 acres to C, giving profit of £1800

3 (i) Finance constraint: $400x + 200y \leqslant 2000$ or $2x + y \leqslant 10$
Staff constraint: $8x + 8y \leqslant 48$ or $x + y \leqslant 6$
Storage constraint: $x + 3y \leqslant 15$
$x \geqslant 0, y \geqslant 0$

(ii) $P = 320x + 240y$

(iii) The objective function is parallel to
$$4x + 3y = 12$$
The required vertex is at the intersection of
$2x + y = 10$ and $x + y = 6$
$\Rightarrow x = 4, y = 2, P = 1760$

(iv)

P	x	y	s_1	s_2	s_3	Value
1	−320	−240	0	0	0	0
0	2	1	1	0	0	10
0	1	1	0	1	0	6
0	1	3	0	0	1	15
1	0	−80	160	0	0	1600
0	1	0.5	0.5	0	0	5
0	0	0.5	−0.5	1	0	1
0	0	2.5	−0.5	0	1	10
1	0	0	80	160	0	1760
0	1	0	1	−1	0	4
0	0	1	−1	2	0	2
0	0	0	2	−5	1	5

After 1st iteration: $x = 5, y = 0$; corresponds to A
After 2nd iteration: $x = 4, y = 2$; corresponds to B

4 Two iterations give $m = 1.625, c = 1.375, l = 0$, covering 87.5 miles. Put the moped in the car and drive for 1.375 hours, covering 55 miles. Then use the moped for the remaining 1.625 hours covering a further 32.5 miles. All of the time is used and $1\frac{11}{12}$ gallons of petrol are used.

5 (i) Max $180x + 90y + 110z$
st $2x + 5y + 3z \leqslant 30$
$4x + y + 2z \leqslant 24$

(ii)

P	x	y	z	s_1	s_2	Value
1	−180	−90	−110	0	0	0
0	2	5	3	1	0	30
0	4	1	2	0	1	24
1	0	−45	−20	0	45	1080
0	0	4.5	2	1	−0.5	18
0	1	0.25	0.5	0	0.25	6
1	0	0	0	10	40	1260
0	0	1	$\frac{4}{9}$	$\frac{2}{9}$	$-\frac{1}{9}$	4
0	1	0	$\frac{7}{18}$	$-\frac{1}{18}$	$\frac{5}{18}$	5

Identification of basic variables + values
$x = 5, y = 4, P = 1260$

(iii) Over two weeks ($x = 3$ and $z = 18$)

(iv) Objective planes are parallel to boundary line.

6 (i) Let w be the number of wardrobes made.
Let u be the number of drawer units made.
Let d be the number of desks made.
Max $80w + 65u + 50d$
st $5w + 3u + 2d < 200$
$4.5w + 5.2u + 3.8d < 200$
$w + 0.75u + 0.5d < 50$

(ii)

M	w	u	d	s_1	s_2	s_3	Value
1	−80	−65	−50	0	0	0	0
0	5	3	2	1	0	0	200
0	4.5	5.2	3.8	0	1	0	200
0	1	0.75	0.5	0	0	1	50
1	0	−17	−18	16	0	0	3200
0	1	0.6	0.4	0.2	0	0	40
0	0	2.5	2	−0.9	1	0	20
0	0	0.15	0.1	−0.2	0	1	10
1	0	5.5	0	7.9	9	0	3380
0	1	0.1	0	0.38	−0.2	0	36
0	0	1.25	1	−0.45	0.5	0	10
0	0	0.025	0	−0.155	−0.05	1	9

$w = 36, u = 0, d = 10$ and $M = 3380$ with
$s_1 = s_2 = 0$ and $s_3 = 9$ (not needed)

(iii) Make 36 wardrobes and 10 desks
Income = £3380
$9\,\text{m}^3$ of storage space spare

7 (i) Let x, y and z be the numbers of litres of X, Y and Z respectively.
Row 1 $\Leftrightarrow$ total profit $P = 10x + 10y + 20z$ (since 40p − 30p = 10p, etc.)
Row 2 $\Leftrightarrow 5x + 2y + 10z \leqslant 10\,000$ (A's availability)
Row 3 $\Leftrightarrow 2x + 4y + 5z \leqslant 12\,000$ (B's availability)
Row 4 $\Leftrightarrow 8x + 3y + 5z \leqslant 8000$ (C's availability)

(ii)

P	x	y	z	s_1	s_2	s_3	Value
1	−10	−10	−20	0	0	0	0
0	5	2	10	1	0	0	10 000
0	2	4	5	0	1	0	12 000
0	8	3	5	0	0	1	8 000
1	0	−6	0	2	0	0	20 000
0	0.5	0.2	1	0.1	0	0	1 000
0	−0.5	3	0	−0.5	1	0	7 000
0	5.5	2	0	−0.5	0	1	3 000
1	16.5	0	0	0.5	0	3	29 000
0	−0.05	0	1	0.15	0	−0.1	700
0	−8.75	0	0	0.25	1	−1.5	2 500
0	2.75	1	0	−0.25	0	0.5	1 500

Make 1500 litres of Y and 700 litres of Z, giving a profit of £290.

(iii) By 16.5p per litre

8 (i) 'He must paint the lower half of each wall in the more expensive paint.'

(ii) 'He has 350 m² of wall to paint.'
The constraint is covered by the requirement to maximise $P = x + y$.

(iii)

P	x	y	s_1	s_2	Value
1	−1	−1	0	0	0
0	1.45	0.95	1	0	400
0	−1	1	0	1	0
1	−2	0	0	1	0
0	2.4	0	1	−0.95	400
0	−1	1	0	1	0
1	0	0	$\frac{5}{6}$	$\frac{5}{24}$	333.33
0	1	0	$\frac{5}{12}$	$-\frac{19}{48}$	166.67
0	0	1	$\frac{5}{12}$	$\frac{29}{48}$	166.67

166.67 m² using more expensive paint and 166.67 m² using less expensive paint. Coverage 333.33 m².

(iv) Coverage = 375 m² (187.5 of each type)

(v) The solution does not maximise the use of the more expensive paint.

Chapter 4

Opening activity Page 58

3 minutes

Activity 4.1 Page 60

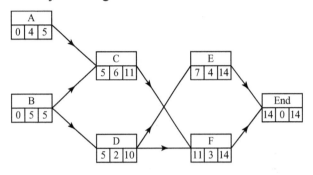

The new minimum completion time is 14 hours.

Exercise 4.1 Page 60

1

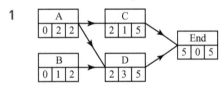

5 days

2

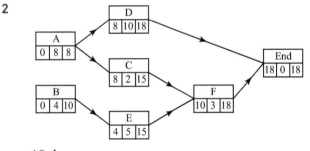

18 days

3

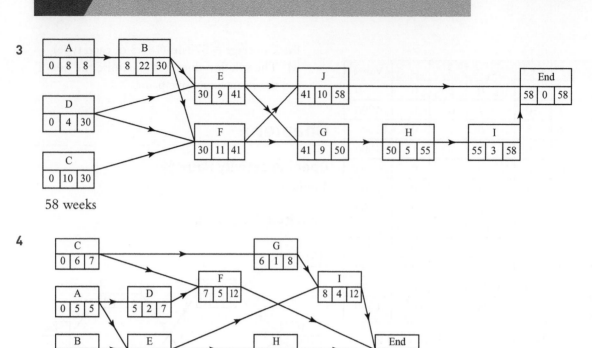

58 weeks

4

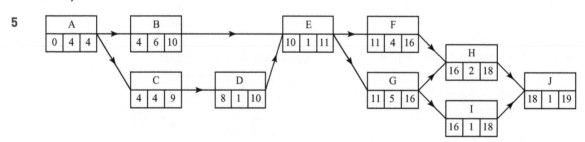

12 days

5

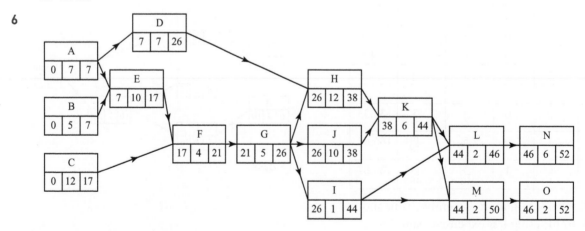

19 weeks

6

52 days: A; E; F; G; H; K; L; N

7

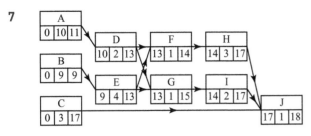

18 weeks: B; E; F; H; J

Activity 4.2 Page 62

As activities A, D and F all have floats of at least 1 hour, increasing their durations has no effect on the minimum completion time. However, each of the critical activities B, C and E will contribute an extra hour, so that the new minimum completion time is 15 + 3 = 18 hours.

Activity A (for example) could overrun by 2 hours, without any effect on the minimum completion time, as the overrunning of B will push the earliest start time for C on by 1 hour, and – because C is a critical activity – its latest finish time will also be pushed on by 1 hour, thereby increasing the float for A.

Exercise 4.2 Page 64

1 (i) Reduce duration of A by 2 days to 6 days at an extra cost of £100.
   (ii) Reduce A by 2 days to 6 days, D by 5 days to 5 days and F by 1 day to 2 days at an extra cost of £410.

2 (i) Critical activities: A, B, F, G, H, I
   (ii) (a) 54 weeks
        (b) E: 1 week; F: 3 weeks; G: 1 week
        (c) £81 000

3 (i) Critical activities: A, D, E, F, I
   (ii) To save 1 day, crash D and E each by 1 day at a cost of £500.
        To save another day crash A and C at a cost of £1600; total cost £2100

4 (i) (a) 30 weeks; B; C; G.
        (b) 40 weeks; A; F.
        (c) Different critical activities.
   (ii) (a) C can complete at 19.5.
            E can complete at 19. So G can start at 19, allowing the project to finish at 31.5.
        (b) Earliest start for C = 10.5.
            Latest finish for C = 31.5.

5 (i) 7.5 days. See figure below.

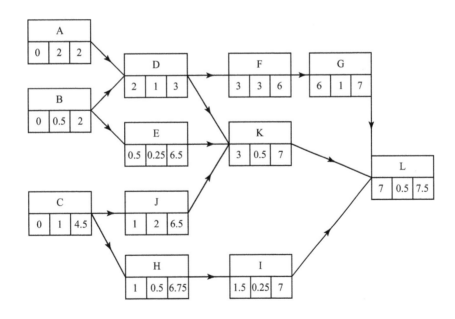

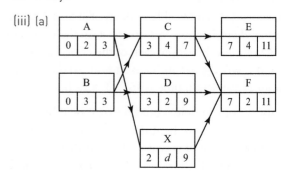

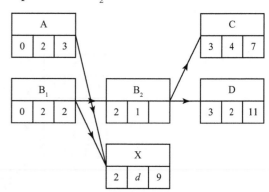

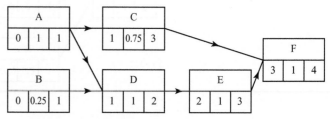

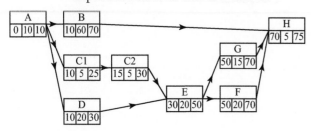

(ii) Add an activity Z of duration 4, dependent on F and preceding K. Critical activities are A, D, F, Z, K and L. Duration becomes 11 days.

6 (i) (ii)

F | 0 | 0.5 | 5
W | 0.5 | 3 | 8
I | 0 | 0.5 | 8
M | 3.5 | 0.5 | 8.5
G | 0 | 2 | 3
B | 2 | 0.5 | 3.5
C | 8.5 | 3 | 11.5
T | 3.5 | 5 | 8.5
D | 0 | 0.5 | 0.5
E | 0.5 | 3 | 3.5
O | 0 | 0.5 | 0.5

(iii) Critical activities: D, O, E, T, C
Minimum time: 11.5 minutes

(iv)
0700	Put toast onto grill
	Dish out cereal
0700.5	Fetch and open milk
0701	Fill kettle
0701.5	Put kettle on to boil
	Eat cereal
0704.5	Put coffee in cup
0705	Make coffee
0705.5	Butter toast
0706	Eat toast
0711	Drink coffee
0714	Finish

7 (i) (a)

A | 0 | 10 | 10
B | 10 | 60 | 70
C1 | 10 | 5 | 40
C2 | 15 | 5 | 45
G | 70 | 15 | 85
H | 85 | 5 | 90
E | 30 | 20 | 65
F | 50 | 20 | 85
D | 10 | 20 | 45

(b) Completion time = 90 minutes
Critical activities: A; B; G; H

(ii) (a) Make G dependent on E only, H dependent on B as well as F and G.

A | 0 | 10 | 10
B | 10 | 60 | 70
H | 70 | 5 | 75
C1 | 10 | 5 | 25
C2 | 15 | 5 | 30
G | 50 | 15 | 70
E | 30 | 20 | 50
F | 50 | 20 | 70
D | 10 | 20 | 30

(b) 75 minutes

8 (i) (a)

A | 0 | 2 | 3
C | 3 | 4 | 7
E | 7 | 4 | 11
B | 0 | 3 | 3
D | 3 | 2 | 9
F | 7 | 2 | 11

(b) Minimum duration = 11 days
Critical activities: B; C; E

(ii) 6 days

(iii) (a)

A | 0 | 2 | 3
C | 3 | 4 | 7
E | 7 | 4 | 11
B | 0 | 3 | 3
D | 3 | 2 | 9
F | 7 | 2 | 11
X | 2 | d | 9

(b) 7 days

(iv) B has to be split into two activities, B_1 and B_2 with X dependent on B_1 and with C and D dependent on B_2.

A | 0 | 2 | 3
C | 3 | 4 | 7
B_1 | 0 | 2 | 2
B_2 | 2 | 1 |
D | 3 | 2 | 11
X | 2 | d | 9

Exercise 4.3 Page 69

1 (i) and (ii)

A | 0 | 1 | 1
C | 1 | 0.75 | 3
F | 3 | 1 | 4
B | 0 | 0.25 | 1
D | 1 | 1 | 2
E | 2 | 1 | 3

Critical activities: A, D, E and F

Minimum duration: 4 days

(iii)

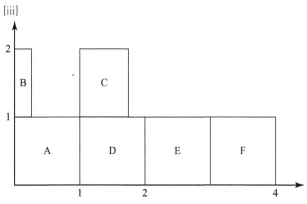

Duration: 5 days

(iv) For example,

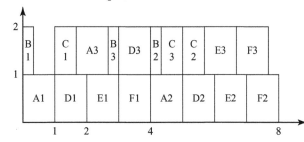

Can be done in 7.5 days however.

2 (i) and (ii)

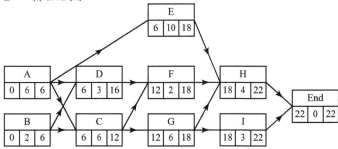

Minimum time to completion = 22 days.
Critical activities are A, C, G and H.

(iii) Latest start time for D is 13 (16 − 3).
Earliest finish time for C is 12.
So the project can be completed in 22 days.

3 (i) (a), (b)

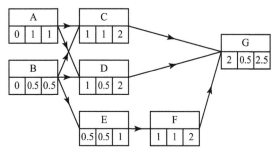

Minimum completion time: 2.5 minutes
Critical activities: A, B, C, E, F, G

(c)

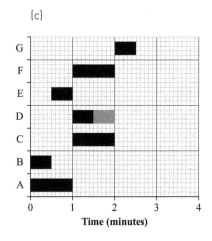

(ii) (a) 8 staff
Activity D (by using 4 staff to shift chairs over 1 minute)
(b) 4 minutes. For example,

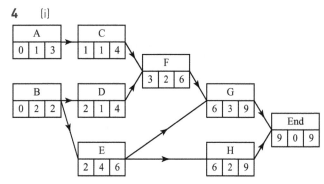

4 (i)

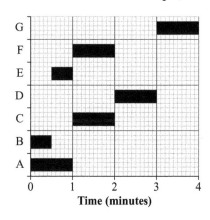

Answers

(ii) 9 days: B; E; G

(iii)

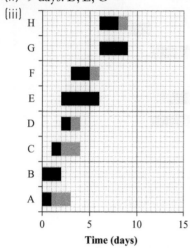

Time (days)

(iv) Start C at beginning of day 4 (i.e. at time 3).
Start F at beginning of day 5
(i.e. at time 4).

5 (i)

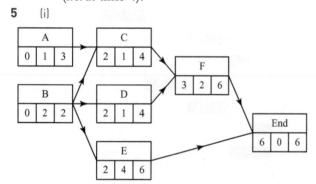

Minimum duration = 6 days.

(ii) Either start C at $t = 2$, D at 3 and F at 4, or start D at 2, C at 3 and F at 4

6 (i) (ii)

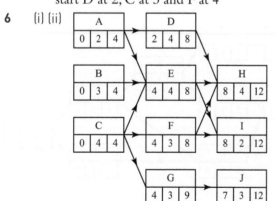

Critical activities: C, E, H
Duration: 12 hours

(iii) $\dfrac{\text{Sum of durations}}{12} = \dfrac{32}{12} > 2$

For example,

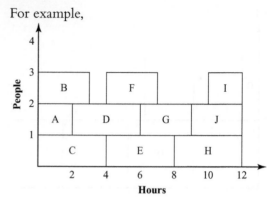

Hours

(iv) Reduce the duration of E by 1 hour, reducing total duration to 11 hours.

7 (i) (ii)

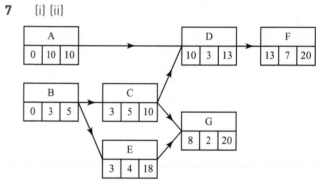

(iii) Minimum completion time is 20 days.
Critical activities are A, D and F.

(iv)

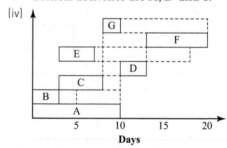

Days

8 (i)

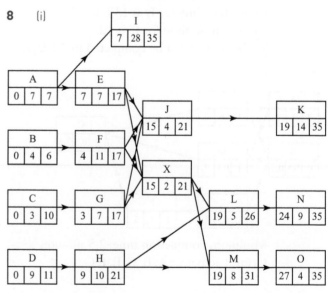

(ii) (a) No, other activities may precede U, affecting the earliest start time.

(b) Yes, 15

(c) No, Q may precede other activities, affecting the latest finish time.

(d) Yes, 21

9 (i)
A	B	C	D	E	F	G	H	I
–	–	–	A,B	B	B,C	D	D,E,F	F

(ii)

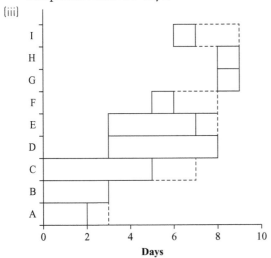

Critical activities are B, D, G, H and minimum completion time is 9 days.

(iii)

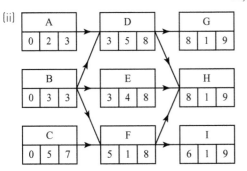

(iv)

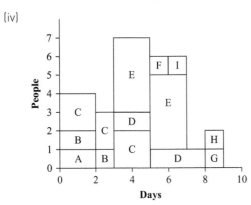

(v) 10 days, E and H

10 (i)

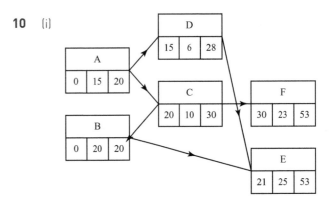

(ii) 53 minutes: B; C; F

(iii) Floats: A–5; D–7; E–7

(iv) B needs to be split into B1 and B2 with A dependent on B1.

54 minutes: B1; A; C; F

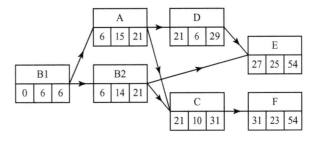

Chapter 5

Opening activity Page 75

ABD, ABCD, ACD, ACBD

Exercise 5.1 Page 78

1 (i) Source is B; sink is E

(ii) 27

2 (i) (a) {S, A, B, C, E}/{D, T}, capacity = 15

(b) {S, A, B, C}/{D, E, T}, capacity = 14

(ii) (a) SA = 2; SB = 6; SC = 7; AD = 2; AB = 0; BD = 5; BE = 1; CB = 0; CE = 7; DT = 7; ET = 8

(b) SA = 2; SB = 5; SC = 7; AD = 2; AB = 0; BD = 5; BE = 0; CB = 0; CE = 7; DT = 7; ET = 7

3 {S}/{ A, B, C, T}, capacity = 22;

{S, A}/{B, C, T}, capacity = 20;

{S, C}/{ A, B, T}, capacity = 39;

{S, A, B}/{C T}, capacity = 25;

{S, A, C}/{B, T}, capacity = 27;

{S, A, B, C}/{T}, capacity = 22

Maximum flow = 20, with flow pattern, for example,

Route	Flow
SABT	10
SCT	7
SCBT	3

4   (i)   B is the source, E is the sink.
    (ii) {A,B,G}/{C,D,E,F,H} has capacity 19;
          {B,C,H}/{A,D,E,F,G} has capacity 34.
    (iii) It is at most 19.

5   $a = 17; b = 2; c = 5; d = 18; e = 7$

6   (i)   {S}/{A,B,C,T} has capacity 14; {S,A}/
          {B,C,T} has capacity 11; {S,B}/{A,C,T} has
          capacity 15; {S,A,B}/{C,T} has capacity 8;
          {S,B,C}/{A,T} has capacity 29; {S,A,B,C}/
          {T} has capacity 13
    (ii) Minimum cut is 8 so maximum flow is 8.
    (iii) Flow is SAT = 3; SBCT
          = 5 so total flow of 8.

Exercise 5.2 Page 84

1   (i)

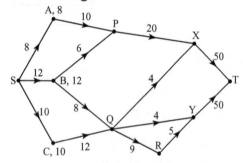

A, 8

    (ii) 27; SA = 8; SB = 12; SC = 7; AP = 8; BP = 6;
          BQ = 6; CQ = 7; PX = 14; QX = 4; QY = 4;
          QR = 5; RY= 5; XT = 18; YT = 9
    (iii) {S,B,C,Q,R}/{A,P,X,Y,T} has value 27.

2   (i)   For example: BA = 8; BC = 7; BE = 12;
          AD = 8; CD = 2; CF = 5; DH = 10; FH = 0;
          FG = 2; FE = 3; HG = 10; GE = 12
    (ii) Total flow (for example BA+BC+BE) = 27.
          This is the same as the capacity of the cut in
          Exercise 5.1 question 1 (ii), so it must be a
          maximal flow.

3   For example:

Route	Flow
SABT	10
SCT	7
SCBT	3

Route	Flow
SABT	10

SCT	5
SCBT	5

4   (i)   Needs SS_1 with capacity 20 and SS_2 with
          capacity 30.
    (ii) Needs T_1T with capacity 25 or more, T_2T
          with capacity 25 or more, and T_3T with
          capacity 20 or more.
    (iii) Cut {S, S_1, S_2, A}/{B, C, D, T_1, T_2, T_3, T} has
          capacity 47. There are many ways of achieving a
          flow of 47, e.g. $T_1T = 25, T_2T = 12, T_3T = 10$.

5   (i)   SABD|CT, capacity = 22
    (ii) Flow = 19 < 22
    (iii) $S \rightarrow A \rightarrow D \rightarrow B \rightarrow C \rightarrow T$ (augmentation
          of 3 possible)
          SA = 10; AD = 5; BD = 4; BC = 8; CT = 13;
          others unchanged.

6   (i)   Maximum flow is 43: BA = 25; BC = 18;
          AH = 12; AE = 13; HC = 0; CD = 18;
          HG = 12; DG = 8; DE = 10; GE = 5; GF = 15;
          EF = 28
    (ii) The value of the cut is 43 so, by the
          maximum flow–minimum cut theorem, the
          flow must be a maximum.

Activity 5.1 Page 86

Route	Flow
SABCT	2
SAT	13
SBCT	6
Total	21

The flows along each of the arcs are shown below.

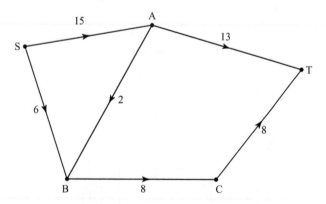

Activity 5.2 Page 86

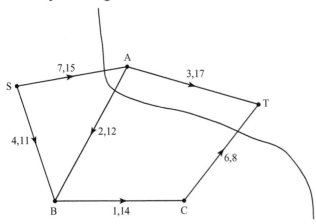

The cut in the diagram has value $15 - 2 + 8 = 21$. Hence the maximum possible flow is 21, as it cannot exceed the cut of 21.

Activity 5.3 Page 86

A flow of 2 is possible along SBAT, to give the following revised network of potential changes.

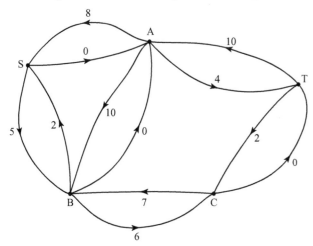

Although it is possible to have a flow along SB, and then a flow along BC, you come to a dead-end at C (as there is no scope for a flow along CT), unless you return to B. At B, the only other route is along BS, which is also a dead-end (as there is no scope for a flow along SA). Therefore you have reached the end of the process. The combined effect of the initial flow and the augmentation is shown in the table below.

Route	Flow
SABCT	4
SAT	11
SBCT	4
SBAT	2

Total	21

The flows along each of the arcs are those already obtained in Activity 5.1.

Exercise 5.3 Page 87

1 Maximum flow = 50.

Route	Flow
SAB_1B_2T	20
SCDT	30

2 (i) Augmentation of 3 available to give a new maximum of 30.
   (ii) Introduce an arc with large capacity from R to the sink T.

3 Maximum flow = 21.

Route	Flow
SABCAT	1
SAT	11
SBCAT	3
SBCT	6

4 (i)

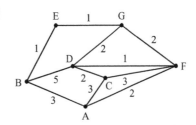

   (ii) BE = 0; BD = 5; BA = 3; EG = 0; DG = 2; DF = 1; DC = 2; AC = 1; AF = 2; GF = 2; CF = 3
   Cut {B,E,D,A,G,C}/{F} has capacity 8, so the flow must be a maximum.

5 (i) Maximum total flow = 19
       e.g. SA = 10; SB = 9; AC = 14; DA = 4; BD = 9; CT = 14; DT = 5
   (ii) S | ABCDT; capacity of cut = 19 so flow must be a maximum.

6 (i)

Route	Flow
SABT	9
SCABT	5
SCBT	5
SCT	8
Total	27

   (ii) {S,C}/{A,B,T} has capacity 27 units so that is the maximum flow.
   (iii) 20, with SC = 11

7 (i) {S,B,D}/{A,B,T} with capacity 24.
The established flow is 18 < 24
Maximum flow is given by:

Route	Flow
SACT	12
SBDT	12
Total	24

(ii) Previous minimum cut now has capacity of
30 as the flow between A and D is from S to
T now. New minimum cut {S,B}/{A,C,D,T}
has capacity 26. Maximum flow is given by:

Route	Flow
SACT	12
SBDT	12
SBDACT	2
Total	26

Chapter 6

Exercise 6.1 Page 93

1 (i) $\begin{pmatrix} -2 & 1 \\ 0 & -3 \\ -1 & 4 \end{pmatrix}$

(ii) 4

(iii) 3

2 (i) Player 1 wins 1
(ii) Player 1 wins 0
(iii) Player 1 wins 2

3 Player 1 chooses option A; player 2 chooses
option B, value is −1

4 (i) Player 1 chooses option B; player 2 chooses
option B
(ii) Player 1 chooses option A; player 2 chooses
option C
(iii) not stable
(iv) not stable

5 (i) Stable; value is 0
(ii) Stable; value is 0
(iii) Not stable
(iv) Not stable

6 The maximum of the row minima is c.
The minimum of the column maxima is also c.

Exercise 6.2 Page 94

1 (i) B & C
(ii) A & D

2 (i) e.g. 1st column dominated; then 1st row;
reduced matrix: (0) [B,B]
(ii) e.g. 2nd row dominated; then columns A, B and
D are dominated; reduced matrix: (0) [A,C]
(iii) No dominated strategies
(iv) No dominated strategies.

3 (i) Player 1 chooses option B;
player 2 chooses option C; value is 0.
(ii) Reduced matrix is (0)

4

	A	B	D	Row minima
A	2	-1	4	-1
B	0	6	2	0
C	4	0	1	0
Column maxima	4	6	4	

5 $x \geqslant 1$

Exercise 6.3 Page 100

1 The value is 2.5

2 (i) Player 1 should never choose option C.
(ii) The value is 1.5

3 (i) 3.4
(ii) 4
(iii) $3\frac{2}{3}$

4 Player 1's strategy: choose A with prob. $\frac{1}{3}$

Player 2's strategy: choose A with prob. $\frac{5}{6}$

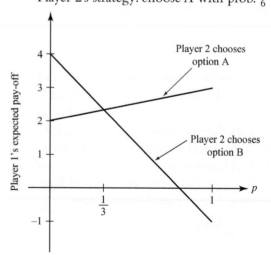

5 Value is 1.8

6 Row C dominates row B, then col. B dominates col. A

Player 1's strategy: choose A with prob. 0.8

Player 2's strategy: choose A with prob. 0.6

Value is 2.6

Exercise 6.4 Page 104

1 Maximise $P = v - 2$, subject to the constraints

$v \leqslant 3p_1 + 4p_2 + 6p_3$,

$v \leqslant 5p_1 + 2p_2 + p_3$,

$p_1 + p_2 + p_3 \leqslant 1$

$(p_1, p_2, p_3 \geqslant 0, v > 0)$

2 The value is 2.5

3 (i) 3.4

(ii) 4

(iii) $3\frac{2}{3}$

4 Maximise $P = v - 4$, subject to the constraints

$v \leqslant 2p_1 + p_2 + 5p_3$,

$v \leqslant 3p_1 + 4p_2 + 2p_3$,

$p_1 + p_2 + p_3 \leqslant 1$

$(p_1 \cdot p_2, \ p_3 \geqslant 0, v > 0)$.

Simplex equations:

$P - v = -4$

$v - 2p_1 - p_2 - 5p_3 + s_1 = 0$

$v - 3p_1 - 4p_2 - 2p_3 + s_2 = 0$

$p_1 + p_2 + p_3 + s_3 = 1$

5

P	v	p	q	s_1	s_2	s_3	Value	
1	−1	0	0	0	0	0	−4	①
0	1	−8	−3	1	0	0	0	②
0	1	−1	−6	0	1	0	0	③
0	0	1	1	0	0	1	1	④

6 $x \geqslant 2$

7 (i) Janet should choose 30 and John should choose 24. Janet wins 6.

(ii) 17

Chapter 7

Opening activity Page 106

1	2	3
2	3	1
3	1	2

Discussion point Page 109

From symmetry about the leading diagonal (from top left to bottom right).

Exercise 7.1 Page 109

1 The result of adding or multiplying modulo m is reduced by a multiple of m until it is in the range 0 to $m - 1$, so they are closed. Addition and multiplication of integers are both commutative and associative, so $+_m$ and $\times_m$ are as well.

2 $x \lozenge (y \lozenge z) = f^{-1}(f(x) \bullet f\{f^{-1}[f(y) \bullet f(z)]\})$

$= f^{-1}(f(x) \bullet [f(y)) \bullet f(z)])$

$= f^{-1}(f(x) \bullet f(y) \bullet f(z))$

$(x \lozenge y) \lozenge z = f^{-1}(f\{f^{-1}[f(x) \bullet f(y)]\} \bullet f(z))$

$= f^{-1}([f(x) \bullet f(y)] \bullet f(z))$

$= f^{-1}(f(x) \bullet f(y) \bullet f(z))$

(i) $xy + 3x + 3y + 6$

(ii) $\dfrac{3xy - 2x - 2y + 1}{2xy - x - y}$

3 (i)

$+_5$	0	1	2	3	4
0	0	1	2	3	4
1	1	2	3	4	0
2	2	3	4	0	1
3	3	4	0	1	2
4	4	0	1	2	3

(ii)

$\times_7$	0	1	2	3	4	5	6
0	0	0	0	0	0	0	0
1	0	1	2	3	4	5	6
2	0	2	4	6	1	3	5
3	0	3	6	2	5	1	4
4	0	4	1	5	2	6	3
5	0	5	3	1	6	4	2
6	0	6	5	4	3	2	1

4 (i)

~	0	1	2	3	4
0	0	1	2	3	4
1	1	0	1	2	3
2	2	1	0	1	2
3	3	2	1	0	1
4	4	3	2	1	0

(ii) Yes

(iii) No

(iv) Yes, 0

(v) Yes, each element is self-inverse.

5 (i) Yes

(ii) Yes

The identity would always be less than any element, so $-\infty$ is the only possible one and that is not a real number. For example, $S = \{x : x \geqslant 0\}$ has identity 0; no element except 0; has an inverse in S.

6 (i) Wednesday, $100 \equiv 2 \pmod 7$

(ii) 03.17 next day; $27.17 \equiv 03.17 \pmod{24}$

(iii) No, no perfect square has 3 as its final digit, i.e. no square $\equiv 3 \pmod{10}$.

7 (i) $x + 2 = y + 4 + 12m \Rightarrow$
$x = y + 2 + 12m$.

(ii) False, e.g. $x = 0, y = 3$.

8 No

9 Both operations are associative.

10 Not associative, $P = (0, 0)$,
$Q = (a_1 - a_2, a_2 - a_1)$,
$I = (0, 0)$, no J.
The argument is invalid since it uses associativity.

Exercise 7.2 Page 111

1 $a, b \in G \Rightarrow e, a^{-1}, b^{-1}, ab \in G$ and $(ab)^{-1} \in G$
So $ab(ab)^{-1} = e$
Pre-multiply by a^{-1} to give $a^{-1}ab(ab)^{-1} = a^{-1}e$
$b(ab)^{-1} = a^{-1}$
Pre-multiply by b^{-1} to give $b^{-1}b(ab)^{-1} = b^{-1}a^{-1}$
$(ab)^{-1} = b^{-1}a^{-1}$

2 Let $ab = c$
$\Leftrightarrow a(ab) = ac$
$\Leftrightarrow a^2b = ac$
$\Leftrightarrow b = ac$
$\Leftrightarrow b^2 = bac$
$\Leftrightarrow e = bac$
$\Leftrightarrow c = bac^2$ post-multiplying by c
$\Leftrightarrow c = ba$
$\Leftrightarrow ab = ba$
e.g. $\begin{pmatrix} 1 & 0 \\ 0 & 1 \end{pmatrix}, \begin{pmatrix} -1 & 0 \\ 0 & -1 \end{pmatrix}, \begin{pmatrix} 0 & 1 \\ 1 & 0 \end{pmatrix}, \begin{pmatrix} 0 & -1 \\ -1 & 0 \end{pmatrix}$

3 Create a Cayley table to determine closure.

$\times_5$	1	2	3	4
1	1	2	3	4
2	2	4	1	3
3	3	1	4	2
4	4	3	2	1

It is closed.
1 is the identity.
Each element has an inverse
(1 is in each row and column).
It is associative because multiplication is associative, so yes, it forms a group.

4 Let $A = \begin{pmatrix} 1 & 0 \\ 0 & 1 \end{pmatrix}$, $B = \begin{pmatrix} 0 & 1 \\ -1 & 0 \end{pmatrix}$,
$C = \begin{pmatrix} -1 & 0 \\ 0 & -1 \end{pmatrix}$, $D = \begin{pmatrix} 0 & -1 \\ 1 & 0 \end{pmatrix}$
Create a Cayley table to determine closure.

$\times$	A	B	C	D
A	A	B	C	D
B	B	C	D	A
C	C	D	A	B
D	D	A	B	C

It is closed.
A is the identity.
Each element has an inverse (A is in each row and column).
Matrix multiplication is associative, so yes, it forms a group.

5 Create a Cayley table to determine closure.

$\times_5$	1	i	−1	−i
1	1	i	−1	−i
i	i	−1	−i	1
−1	−1	−i	1	i
−i	−i	1	i	−1

It is closed.
1 is the identity.
Each element has an inverse
(1 is in each row and column).
Multiplication is associative, so yes, it forms a group.

6 $x = a^{-1}b$, $y = ab$

7 (ii), (iii), (iv) are groups,
(iii) is abelian.

Activity 7.1 Page 112

	1	2	4	8
1	1	2	4	8
2	2	4	8	1
4	4	8	1	2
8	8	1	2	4

Discussion point Page 112

For any two elements a and b,
$a = g^r$ and $b = g^s$ for some integers r and s, where g is a generator of the group. Then $ab = g^r g^s = g^{r+s} = g^s g^r$ (by associativity) $= ba$.

Thus all cyclic groups are abelian.

Exercise 7.3 Page 113

1 Let ab have period n.
$\Leftrightarrow (ab)^n = e$
$\Leftrightarrow abab \ldots ab = e$
$\Leftrightarrow a^{-1}abab \ldots aba = a^{-1}ea$ (pre-multiply by the inverse of a and post-multiply by a)
$\Leftrightarrow bab \ldots aba = e$
$\Leftrightarrow (ba)^n = e$ so ba has period n, and hence the result follows.

2

$\times_7$	4	4^2	4^3
4	4^2	4^3	4
4^2	4^3	4	4^2
4^3	4	4^2	4^3

3 $(a)^n = e$
$\Leftrightarrow (a^{-1})^n (a)^n = (a^{-1})^n e$
$\Leftrightarrow e = (a^{-1})^n$
Hence result.

4 (i) Let $C_5 = \{e, x, x^2, x^3, x^4\}$
Clearly x generates the group with $x^5 = e$.
x^2: $(x^2)^1 = x^2$; $(x^2)^2 = x^4$;
$(x^2)^3 = x$; $(x^2)^4 = x^3$;
$(x^2)^5 = e \Rightarrow x^2$ is a generator.
x^3: $(x^3)^1 = x^3$; $(x^3)^2 = x$;
$(x^3)^3 = x^4$; $(x^3)^4 = x^2$;
$(x^3)^5 = e \Rightarrow x^3$ is a generator.
x^4: $(x^4)^1 = x^4$; $(x^4)^2 = x^3$;
$(x^4)^3 = x^2$; $(x^4)^4 = x$;
$(x^4)^5 = e \Rightarrow x^4$ is a generator.
Let $C_6 = \{e, \gamma, \gamma^2, \gamma^3, \gamma^4, \gamma^5\}$
Clearly γ generates the group with $\gamma^6 = e$.
γ^2: $(\gamma^2)^1 = \gamma^2$; $(\gamma^2)^2 = \gamma^4$;
$(\gamma^2)^3 = e$; $(\gamma^2)^4 = \gamma^2$;
$(\gamma^2)^5 = \gamma^4 \Rightarrow \gamma^2$ is not a generator of the group as γ, γ^3 and γ^5 are not generated.

(ii) True, false
(iii) k, n have no common factor.

5 9 and 20 have no common factor, so rotation through $\frac{9\pi}{10}$ generates the whole set of rotations. Other possible values of n (<20) are 1, 3, 7, 11, 13, 17, 19.

6 (i) $3 * (9 * 11) = 3 * 3 = 9$;
$(3 * 9) * 11 = 11 * 11 = 9 \Rightarrow$ associative
(ii) Construct a Cayley table.

*	1	3	9	11
1	1	3	9	11
3	3	9	11	1
9	9	11	1	3
11	11	1	3	9

The table shows closure, 1 is the identity, each element has an inverse, and associativity has been assumed.
(iii) Periods of 1, 3, 9, 11 are 1, 4, 2, 4.
(iv) $3^1 = 3$; $3^2 = 9$; $3^3 = 11$;
$3^4 = 1 \Rightarrow 3$ generates the group, so it is cyclic.

Activity 7.2 Page 114

A Cayley table for this group is shown below.

	e	a	b	c
e	e	a	b	c
a	a	e	c	b
b	b	c	e	a
c	c	b	a	e

This is the Klein 4-group again.

Activity 7.3 Page 115

The Cayley table for the group is as follows.

	e	a	b	c	d	f
e	e	a	b	c	d	f
a	a	e	d	f	b	c
b	b	c	e	a	f	d
c	c	b	f	d	e	a
d	d	f	a	e	c	b
f	f	d	c	b	a	e

(i) No: for example, $ab = d$ but $ba = c$.
(ii) $a^2 = e$, so a is of period 2; $b^2 = e$, so b is of period 2;
$c^2 = d$, $c^3 = c(c^2) = cd = e$, so c is of period 3.
$c(c^2) = c(c)(c) = (c^2)c$, by associativity, so that $cd = dc$ (even though the group isn't abelian).
$d^2 = c$, $d^3 = d(d^2) = dc = e$, so d is of period 3;
$f^2 = e$, so f is of period 2.
Thus the elements e, a, b, c, d and f have periods 1, 2, 2, 3, 3, 2, respectively.

Exercise 7.4 Page 116

1

		First transformation						
	Followed by	I	R	H	S	X	Y	A	B
Second transformation	I	I	R	H	S	X	Y	A	B
	R	R	H	S	I	A	B	Y	X
	H	H	S	I	R	Y	X	B	A
	S	S	I	R	H	B	A	X	Y
	X	X	B	Y	A	I	H	S	R
	Y	Y	A	X	B	H	I	R	S
	A	A	X	B	Y	R	S	I	H
	B	B	Y	A	X	S	R	H	I

It is closed, there is an identity, I, and each element has an inverse (I is in each row and column), and it is associative.

2 (i)

$\times_{15}$	1	4	7	13
1	1	4	7	13
4	4	1	13	7
7	7	13	4	1
13	13	7	1	4

Closed; identity is 1; inverse as 1 is in each row/column; associative; commutative (symmetry about leading diagonal)

$\times_{15}$	3	6	9	12
3	9	3	12	6
6	3	6	9	12
9	12	9	6	3
12	6	12	3	9

Closed; identity is 6; inverse as 6 is in each row/column; associative; commutative (symmetry about leading diagonal).

$\times_{15}$	2	5	8	14
2	4	10	1	13
5				
8				
14				

Not closed, so not a group.

(ii)

$\times_8$	1	2	3	4	5	6	7
1	1	2	3	4	5	6	7
2	2	4	6	0	2	4	6
3	3	6	1	4	7	2	5
4	4	0	4	0	4	0	4
5	5	2	7	4	1	6	3
6	6	4	2	0	6	4	2
7	7	6	5	4	3	2	1

Not closed.
For example, $\{1, 3\}$,
$\{1, 3, 5, 7\}$.

3 (i) Let $f(x) = x$; $g(x) = -\dfrac{1}{x}$;

$h(x) = \dfrac{x-1}{x+1}$; $k(x) = \dfrac{1+x}{1-x}$

comp	f	g	h	k
f	f	g	h	k
g	g	f	k	h
h	h	k	g	f
k	k	h	f	g

Closed; identity is f; inverse as f is in each row/column; associative; commutative (symmetry about leading diagonal)

(ii) The Klein 4-group has the identity along the leading diagonal and this group does not, so not the Klein 4-group.

(iii) It is a cyclic group.

4 (i) G is closed as the Cayley table is:

	P	Q	R	S
P	P	Q	R	S
Q	Q	P	S	R
R	R	S	P	Q
S	S	R	Q	P

Showing closure, identity is P, each element has an inverse, itself, and associativity may be assumed, so G is a group under matrix multiplication.

(ii) 1, 2, 2, 2 respectively

5 (i) $H = R^2$, $S = R^3$, $Y = R^2X$, $A = RX$, $B = R^3X$; $I = X^2$ so all elements can be generated by products of R and X.

(ii) R is a rotation of 90°, so $R^4 = I$; X is a reflection, so $X^2 = I$; so $R^4 = X^2 = I$; $XR^3 = (XR)R^2 = BH = A = RX$ from Cayley table in question 1, so $RX = XR^3$.

6 Let R = rotation of $\dfrac{360°}{n}$ about the centre of the polygon $\Rightarrow$ all rotations are $R^2, R^3 \dots R^{n-1}; I = X^2 = R^n$; let X be a reflection in the line of symmetry through one vertex, so reflection in the lines of symmetry through the other vertices will be a combination of X followed by a rotation through a multiple of $\dfrac{360°}{n}$ and hence D_n is generated by X and $A^2 = RXRX = I$

$\Rightarrow RXRX^2 = X \Rightarrow RXR = X \Rightarrow RXRR^{n-1} = XR^{n-1}$
$\Rightarrow RXR^n = XR^{n-1} \Rightarrow RX = XR^{n-1}$

Activity 7.4 Page 117

You have already seen that the elements a, b and f have period 2. This means that $\{e,a\}$, $\{e,b\}$ and $\{e,f\}$ are subgroups.

Also, you saw that $c^2 = d$ and $c^3 = e$, so that $\{e, c, d\}$ is a subgroup as well.

Activity 7.5 Page 117

Referring to the Cayley table for the Klein 4-group (Table 7.7), you see that $a^2 = e, b^2 = e, c^2 = e.$

Thus the Klein 4-group has three elements of period 2, and its proper subgroups are $\{e\}$, $\{e,a\}$, $\{e,b\}$, and $\{e,c\}$.

	e	a	b	c
e	e	a	b	c
a	a	e	c	b
b	b	c	e	a
c	c	b	a	e

Discussion point Page 117

If G is a group of prime order, it follows from Lagrange's theorem that the only proper subgroup of G is $\{e\}$. Therefore, as any element other than e will generate a subgroup, it must generate G itself. Hence G is cyclic.

Exercise 7.5 Page 117

1 (i) There are 4 lines of symmetry so 4 reflections, rotational symmetry is order 4 so 4 rotations, one of which is the identity. Group has 8 elements, so order is 8.

		First transformation							
Followed by		I	R	H	S	X	Y	A	B
Second transformation	I	I	R	H	S	X	Y	A	B
	R	R	H	S	I	A	B	Y	X
	H	H	S	I	R	Y	X	B	A
	S	S	I	R	H	B	A	X	Y
	X	X	B	Y	A	I	H	S	R
	Y	Y	A	X	B	H	I	R	S
	A	A	X	B	Y	R	S	I	H
	B	B	Y	A	X	S	R	H	I

(ii) R, S have period 4; H, X, Y, A and B have period 2. I is the identity.

(iii) $\{I, H\}, \{I, X\}, \{I,Y\}, \{I,A\}, \{I, B\}, \{I, R, H, S\}, \{I, H, X, Y\}, \{I, H, A, B\}$

(iv) $\{I, H\}, \{I, X\}, \{I,Y\}, \{I,A\}, \{I, B\}, \{I, R, H, S\}$

2 (i) $g \in H$ and $g \in K \Rightarrow$ $g \in H \cap K \Rightarrow g^2 \in H$ and $g^2 \in K \Rightarrow g^2 \in H \cap K$ and so on; therefore $H \cap K$ is a subgroup of G.

(ii) For example,
G = (integers, +),
H = {multiples of 2},
K = {multiples of 3}, $H \cup K$ is not closed since $2 + 3 = 5$ is not a member of it.

3 $\mathbb{Z}_6$ has order 6, so there may be subgroups of order 2 or 3. $\{0, 3\}$, $\{0, 2, 4\}$, $\{0\}$.

4 (i) $s_1, s_2 \in S \Rightarrow s_1 s_2 \in S$;
$t_1, t_2 \in T \Rightarrow t_1 t_2 \in T$;
$s_1 t_1, s_2 t_2 \in ST; \Rightarrow (s_1 t_1)(s_2 t_2)$
$= s_1(t_1 s_2) t_2$
$= s_1 s_2 t_1 t_2) \in ST$ so closed.
$e \in S, T \Rightarrow e \in ST$;
ST is associative since G is associative;
$(s_1 t_1)(s_1 t_1)^{-1}$
$= s_1 t_1 t_1^{-1} s_1^{-1} = s_1(t_1 t_1^{-1})s_1^{-1} = s_1 s_1^{-1} = e$
and vice versa, so ST is a group.

(ii) $s_1 t_1 = s_2 t_2 \Rightarrow s_1^{-1} s_1 t_1 t_2^{-1}$
$= s_1^{-1} s_2 t_2 t_2^{-1} \Rightarrow t_1 t_2^{-1}$
$= s_1^{-1} s_2$. Since the only element in both sets is
e, $t_1 t_2^{-1} = e$, $s_1^{-1} s_2 = e \Rightarrow$
$s_1 = s_2, t_1 = t_2$.

(iii) If S contains k elements and T contains m elements, by the product rule for counting, ST contains km elements.

(iv) $\{1, 5, 13, 17, 25, 29\}$; 5 is a generator.

(v) $\{1, 17, 19, 35\}$; $17^2 \equiv 19^2 \equiv 35^2 \equiv 1 \pmod{36}$; thus the subgroup cannot be generated from a single element, and is therefore not cyclic.

5 (i) $(A^2B)(AB) = A^2(BA)B$
$= A^2 A^3 BB = AA^4 B^2 = A$; $(AB)(A^2B) = A(AB) AB$
$= AA^2 BAB = A^4 A^3 BB$
$= A^3 B^2 = A^3$

(ii) B, I, A^2B

(iii) 1, 4, 2, 4, 2, 2, 2, 2 respectively

(iv)

	I	A²	B	A²B
I	I	A²	B	A²B
A²	A²	I	A²B	B
B	B	A²B	I	A²
A²B	A²B	B	A²	I

The Cayley table shows closure, the existence of the identity and an inverse for each element. G is associative, so the subgroup is also. Thus the set is a subgroup of G.

(v) $\{I, A, A^2, A^3\}$,
$\{I, A^2, AB, A^3B\}$

(vi) Only $\{I, A, A^2, A^3\}$ is cyclic.

6 (i) $0 < m < k \Rightarrow 1 < 2^m < 2^k \Rightarrow 0 < 2^m - 1 < 2^k - 1 \Rightarrow$
$0 < 2^m - 1 < 2^k + 1 = p$
Hence $2^m - 1 \not\equiv 0$, and therefore $2^m \not\equiv 1$ (mod p).

(ii) $k < m < 2k \Rightarrow 0 < 2k - m < k$ (considering the number line from k to $2k$, the distance $2k - m$ is shorter than $2k - k$); then (i) $\Rightarrow 2^{2k-m} - 1 \not\equiv 0$ (replacing m with $2k - m$), so that $2^{2k-m} \not\equiv 1$. Suppose that $2^m \equiv 1$. Now $2^{2k} = (p-1)^2 = p^2 - 2p + 1 \equiv 1$. Hence $2^{2k} \equiv 2^m$; i.e. $2^{2k} = 2^m + np$ (where $n \in \mathbb{Z}^+$), so that $2^{2k-m} = 1 + \frac{np}{2^m}$; as $2^{2k-m} \in \mathbb{Z}^+$, it follows that $\frac{np}{2^m}$

$\in \mathbb{Z}^+$, and p prime $\Rightarrow \frac{n}{2^m} \in \mathbb{Z}^+ \Rightarrow 2^{2k-m} \equiv 1$, contradicting $2^{2k-m} \not\equiv 1$. Hence $2^m \not\equiv 1$ (mod p).

(iii) Let the period of 2 be N where $N \neq 0$. Then $2^N \equiv 1$ (the identity element of G). From (i) and (ii), $0 < N < k$ and $k < N < 2k$ are not possible. Also, $2^k = p - 1 \not\equiv 1$. So, as $2^{2k} \equiv 1$ (from (ii)), it follows that $N = 2k$

(iv) By Lagrange's theorem, the order of the subgroup generated by 2 divides the order of G; i.e. $2k$ divides $p - 1$, so that $2^k = 2kM$, where $M \in \mathbb{Z}^+$. Thus the only prime factors of $2kM$ are 2, and it follows that k is a power of 2.

Exercise 7.6 Page 119

1 $a, b \in A$, an abelian group, and $c, d \in B$, a non-abelian group. $ab = ba$, for all $a, b \in A$, but $cd \neq dc$, for some c and d, so A and B cannot be isomorphic.

2 For any $x, y \in \mathbb{R}^+$, there is $\log x, \log y \in \mathbb{R}$, and $xy \in \mathbb{R}^+$ corresponds to $\log x + \log y \in \mathbb{R}$.

3 Let $\begin{pmatrix} p & 0 \\ 0 & p \end{pmatrix}$ correspond with $\begin{pmatrix} 1 & 0 \\ p & 1 \end{pmatrix}$ and $\begin{pmatrix} q & 0 \\ 0 & q \end{pmatrix}$ correspond with $\begin{pmatrix} 1 & 0 \\ q & 1 \end{pmatrix}$; $\begin{pmatrix} p & 0 \\ 0 & p \end{pmatrix} + \begin{pmatrix} q & 0 \\ 0 & q \end{pmatrix} = \begin{pmatrix} p+q & 0 \\ 0 & p+q \end{pmatrix}$,
$\begin{pmatrix} 1 & 0 \\ p & 1 \end{pmatrix}\begin{pmatrix} 1 & 0 \\ q & 1 \end{pmatrix} = \begin{pmatrix} 1 & 1 \\ p+q & 0 \end{pmatrix}$

So $\begin{pmatrix} p+q & 0 \\ 0 & p+q \end{pmatrix}$ corresponds to $\begin{pmatrix} 1 & 0 \\ p+q & 1 \end{pmatrix}$ and so the two groups are isomorphic

4 (i) $ab = a \Rightarrow b = e$, $ab = b \Rightarrow a = e$; both are impossible since a, b and e are distinct.

(ii) $ab \neq a$ or b shown in (i), similarly $ba \neq a$ or b; so ab, ba can each only be e or c. The four possibilities are the possible combinations of e and c for ab and ba.

(iii) Case (b): The second row cannot be completed as c cannot be in either position (it appears twice in a row or column)

	e	a	b	c
e	e	a	b	c
a	a		e	
b	b	c		
c	c			

Case (c) The third row cannot be completed as c cannot be in either position (it appears twice in a column)

	e	a	b	c
e	e	a	b	c
a	a		c	
b	b	e		
c	c			

Case (a)

	e	a	b	c
e	e	a	b	c
a	a	c	e	b
b	b	e	c	a
c	c	b	a	e

Case (d)

	e	a	b	c
e	e	a	b	c
a	a	e	c	b
b	b	c	e	a
c	c	b	a	e

	e	a	b	c
e	e	a	b	c
a	a	e	c	b
b	b	c	a	e
c	c	b	e	a

	e	a	b	c
e	e	a	b	c
a	a	b	c	e
b	b	c	e	a
c	c	e	a	b

5 $\mathbb{Z}_6$ has Cayley table

	0	1	2	3	4	5
0	0	1	2	3	4	5
1	1	2	3	4	5	0
2	2	3	4	5	0	1
3	3	4	5	0	1	2
4	4	5	0	1	2	3
5	5	0	1	2	3	4

D_3 has Cayley table

	I	P	Q	U	V	W
I	I	P	Q	U	V	W
P	P	Q	I	W	U	V
Q	Q	I	P	V	W	U
U	U	V	W	I	P	Q
V	V	W	U	Q	I	P
W	W	U	V	P	Q	I

Comparing the leading diagonals, there are 3 elements with period 2 in D_3 and only one in $\mathbb{Z}_6$, so they cannot be isomorphic.

6 (i) Let a group of order 2 consist of the two elements $\{e, a\}$

Then $ee = e, ea = ae = a, aa = e$. There is no other possible structure, so all groups of order 2 are isomorphic.

(ii)
e a b
a b e
b e a

Since there is only one way of completing the Cayley table, all groups of order 3 are isomorphic.

e.g. $(\{1, \omega, \omega^2\}, \times)$

7 (i)

$\times$	1	-1	i	$-i$
1	1	-1	i	$-i$
-1	-1	1	$-i$	i
i	i	$-i$	-1	1
$-i$	$-i$	i	1	-1

The table shows: Closure; an identity, 1; each element has an inverse; multiplication of complex numbers is associative

(ii)

	e	f	g	h
e	e	f	g	h
f	f	e	h	g
g	g	h	f	e
h	h	g	e	f

The table shows: Closure; an identity, e; each element has an inverse; composition of functions is associative

(iii)

×₅	1	2	3	4
1	1	2	3	4
2	2	4	1	3
3	3	1	4	2
4	4	3	2	1

The table shows: Closure; an identity, 1; each element has an inverse; multiplication of real numbers is associative

(iv)

×₈	1	3	5	7
1	1	3	5	7
3	3	1	7	5
5	5	7	1	3
7	7	5	3	1

The table shows: Closure; an identity, 1; each element has an inverse; multiplication of real numbers is associative

(v) Let $I = \begin{pmatrix} 1 & 0 \\ 0 & 1 \end{pmatrix}$, $P = \begin{pmatrix} 1 & 0 \\ 0 & -1 \end{pmatrix}$,

$Q = \begin{pmatrix} -1 & 0 \\ 0 & 1 \end{pmatrix}$, $R = \begin{pmatrix} -1 & 0 \\ 0 & -1 \end{pmatrix}$

	I	P	Q	R
I	I	P	Q	R
P	P	I	R	Q
Q	Q	R	I	P
R	R	Q	P	I

The table shows: Closure; an identity, I; each element has an inverse; multiplication of matrices is associative

(vi) Let $I = \begin{pmatrix} 1 & 0 \\ 0 & 1 \end{pmatrix}$, $X = \begin{pmatrix} 0 & 1 \\ -1 & 0 \end{pmatrix}$,

$Y = \begin{pmatrix} -1 & 0 \\ 0 & -1 \end{pmatrix}$, $Z = \begin{pmatrix} 0 & -1 \\ 1 & 0 \end{pmatrix}$

	I	X	Y	Z
I	I	X	Y	Z
X	X	Y	Z	I
Y	Y	Z	I	X
Z	Z	I	X	Y

The table shows: Closure; an identity, I; each element has an inverse; multiplication of matrices is associative

(i), (ii), (iii), (vi) are isomorphic; (iv), (v) are isomorphic

8 (For brevity the elements of each group are called e, a, b, c respectively.)
Permutations and matrices: periods 1, 2, 2, 2; non-cyclic; proper subgroups $\{e\}, \{e, a\}, \{e, b\}, \{e, c\}$
Complex numbers: periods 1, 2, 4, 4; cyclic; proper subgroups $\{e\}, \{e, a\}$
All groups of order 4 are isomorphic to either the Klein 4-group or the cyclic group of order 4, so there can be no others.

9 (i)

	1	3	7	9	11	13	17	19
1	1	3	7	9	11	13	17	19
3	3	9	1	7	13	19	11	17
7	7	1	9	3	17	11	19	13
9	9	7	3	1	19	17	13	11
11	11	13	17	19	1	3	7	9
13	13	19	11	17	3	9	1	7
17	17	11	19	13	7	1	9	3
19	19	17	13	11	9	7	3	1

(ii)

x	1	3	7	9	11	13	17	19
x^{-1}	1	7	3	9	11	17	13	19

(iii)

x	1	3	7	9	11	13	17	19
Period	1	4	4	2	2	4	4	2

(iv) $\{1\}, \{1, 9\}, \{1, 11\}$,
$\{1, 19\}, \{1, 3, 7, 9\}$,
$\{1, 9, 13, 17\}$,
$\{1, 9, 11, 19\}, G$
$\{1, 9\}, \{1, 11\}, \{1, 19\}$ are isomorphic

(v) (a) 0 has no inverse, so J is not a group.
(b) K is closed and inverses of 0, 1, 2, 3, 4, 5, 6, 7 are 0, 7, 6, 5, 4, 3, 2, 1 respectively, so K is a group.
Different pattern (two self-inverse elements) K is not isomorphic to G.